NEWT
IN THE WORLD OF
TARZAN

NEWT
IN THE WORLD OF
TARZAN

Marshall B. Allen, Jr.

Augusta, Georgia

Newt in the World of Tarzan
By Marshall B. Allen, Jr.
A Harbor House Book/2007

Copyright 2007 by Marshall B. Allen, Jr.

All rights reserved. No part of this book may be reproduced or transmitted in any form or by any means, electronic or mechanical, including photocopying, recording, or by any information storage and retrieval system, without permission in writing from the publisher.

For information address:
    HARBOR HOUSE
    111 TENTH STREET
    AUGUSTA, GA 30901

Jacket and book design by Nathan Elliott
Photo Courtesy of Newton Fiveash, Jr.
Library of Congress Cataloging-in-Publication Data

Allen, Marshall B., 1927-
  Newt in the world of Tarzan / by Marshall B. Allen, Jr.
    p. cm.
  ISBN-13: 978-1-891799-89-1
  I. Title.
  PS3601.L4326N49 2007
  813'.6--dc22
                        2007000471

Printed in the United States of America
10 9 8 7 6 5 4 3 2 1

Dedication

Newt in the World of Tarzan is dedicated to the man whose escapades it relates. Newt is a great man. Having started life with a fifth-grade education, he has made the most of his 95 years. Almost everything related here is based on facts that Newt has revealed through long talks with me and his friends, of whom he has many. Having come from an earlier generation, Newt is still dedicated to the "Old South," but his experiences have been so varied that his Old South mindset hardly comes through within these pages. Here's wishing Newt the best always.

Preface

Newton Fiveash, Jr. is one of the most fascinating persons I have had the pleasure of knowing. At 95 years of age, he has experienced five different lives. Born and raised on farms in South Georgia, he left home at 15 years of age and worked as a "gofer" in Florida for seven years. During that time, among many other jobs he held, he worked with the movie companies making Tarzan films and was used as a stand-in when needed for some of the jungle escapades, particularly in *Tarzan Finds a Son*.

In the height of the Depression, Newt had to find a paying job and went to work for the Florida Motor Lines, later taken over by the Greyhound Bus System. When he retired, he was given credit for thirty-eight years as a bus driver, even though five of those years were spent in the U.S. Navy. After his Navy discharge, one day he was driving his bus along a highway and saw a sign for a nudist camp. At the end of his run, he called back to visit. He was so struck by the lifestyle that he talked his wife into joining and eventually opened his own camp, the subject of *The Naked Bus Driver*.

Starting from the time he was in the Navy, Newt was plagued with severe back pain, which began to radiate into his lower extremities. He was admitted to a local hospital where I operated on him and came to know him well.

Because Newt is such and outstanding storyteller, I have been able to put together some of his yarns. This book relates some of his experiences in the forest and river where the first Tarzan movies were being filmed. While many parts of the book are fictionalized, they are all based on actual events. I hope you have as much fun reading about Newt's experiences as I have had writing about them.

1

"**NEWT, AFTER YOU** and Brooks feed the cows and mules and milk in the morning, I want you to hoe around the okra and tomatoes. It'll probably take you at least two days, maybe three, to get the grass cleaned out. After you get the weeds out of there, take the mules over and plow between the rows of corn in the back field. While you're cleaning up the fields over here, Preacher and Virgil'll be working over at the other place. There's also pole beans and butter beans and some tomatoes to be picked, but Wesley and Grover and the girls can do that. Next week, we'll need to be getting the ground ready for planting more corn. But we'll talk about that next weekend."

Daddy was giving Brooks and me orders, what he wanted—no, expected—us to do on the farm while he was working on the railroad. It was Sunday evening, just two days after the end of the school year. I was 14 years old, going on 15, at the time, and I had just finished the fifth grade. Brook was two years younger. Summer vacation, if that's what you call it, was just starting, and it was the beginning of a scorcher.

I was the oldest of seven children, one of whom was adopted. For the last two years, Brooks and I had managed the farms, or at least, we had worked on them every weekday when school was out. We had to make sure that Daddy's orders were carried out. Preacher, a black man, and his wife and three children lived in a little cabin behind our house. Virgil and his family lived on our other farm outside of town, but Preacher and Virgil worked together as a team, usually on the farm where Virgil lived. It was about three miles away, but easy to get to by horse and wagon, or on foot, which was usually Preacher's way.

During the school year, I had to feed and milk the cows, feed the mules and slop the hogs every day before breakfast and after school. During the summer, Brooks and I had to clean the barnyard in the morning and then do our share of the plowing and planting or cutting hay, just whatever other work there was to be done on the farm. In the fall, after school started, the sugarcane had to be squeezed, but Preacher did that during the week and sometimes on Saturdays when we didn't have to butcher hogs. We killed hogs after the weather started cooling off. Preacher always helped me do that. In fact, you might say that I helped Preacher since he knew a lot more about butchering hogs than I did, at least until the last year I was at home, and to tell the truth, I learned everything I knew about butchering hogs from Preacher. After the hogs were killed and dressed, we salted the meat and hung the hams in the smokehouse. There were oak trees to be cut for wood to smoke the meat. There was always plenty of work for every one of us.

Daddy was a conductor on the South Georgia Railroad, running from Hawkinsville, through Ashburn and Sylvester and on to Camilla. We called it the "GAS Line." I guess in these days you would call that an acronym for Georgia, Ashburn to Sylvester. Daddy rode the trains almost every week day, but he was off on weekends, so he was at home then, or at least he was in town. He always spent lots of time down around the courthouse making business deals.

Daddy was a little man, but he was a real "go-getter." I heard some people say he had a Napoleonic personality, but I didn't know what that meant. He could always think of lots of work for us to do. He had a reputation across town as well as at home for squeezing every penny until it squealed.

A few months after I was born, Daddy bought the farm on the south side of Ashburn, and two years later, he bought the other farm out west of town. We lived on the first place he bought, but when I was 11 years old, he expected me to do my work and keep watch on the farmhands as well. For our work on the farm, Brooks and I were given fifty cents a week allowance. That seemed like a lot of money at the time, but then we did a lot of work. It was still only a little part of what Daddy was paying the field hands, that we were supposed to be watching. Daddy always gave Brooks and me our orders on Friday and Sunday evenings. At this week's meeting, he was trying to get the summer's work off to a good start.

"Yes, Sir," I answered after each order, sounding as though I were in the Navy, although I had no military experience and didn't know what I was doing. I just knew that those words sounded good and I was taking orders from my boss.

It was just after my third "Yes, Sir" that it popped into my mind that now was the time to ask Daddy that question I'd been thinking about for weeks. So I took in a deep breath and blurted out, "Daddy, do you think I might be able to go down to Okena for a little vacation the last week of the summer? I'd like to visit Aunt Kate and my cousins."

"Sure. That's a good idea. You can do that. I'd like for you to get to know your cousins better."

At the time, I was looking forward to my fifteenth birthday in August, and like many young'uns, I was thinking about getting to see how the world looked outside the town of Ashburn. The only places I remembered having ever seen were in South Georgia. I had been to the home of my grandparents down near the Okefenokee Swamp, to Tifton to see the fair and once to Waycross to see a ball game. I had ridden on Daddy's train through the farms where they grew peanuts, watermelons and cantaloupes, and always turnips, collard greens, peas, string beans and corn, lots of corn.

During summer months, the trains, headed south, stopped at nearly every sidetrack to leave empty box cars. On the way back the cars were picked up when they were full of cantaloupes and watermelons to be hauled up north. I still didn't know much of anything about other parts of the world except what I had read about in geography books. I was getting the itch to see more. I'd heard a few things about goings-on in Florida, and I had even heard about wild parties the kids up north had every once in a while.

WHEN I ASKED about the vacation, Daddy must have been thinking about work on the farms, and maybe what I wanted didn't seem very important, but maybe he thought that I might work harder if he promised me a vacation. Anyway, it must have gone through his mind that I had three months to save enough money to pay for my bus ticket to Okena and back. However now, years later, my guess is that he really didn't think much about it, certainly not like I did. He probably just thought that, since the time when I wanted to visit was three months away, it would be easy to work out by then.

After Daddy got me off his back, he went on talking about what needed to be done on the farm during the next week. He repeated some of his orders. "On Thursday, you and Brooks need to take the mules and middle busters out and plow between the rows of corn in the back field. The grass is getting ahead of us, The ground needs to be broken, and the grass turned under."

So, Brooks and I went right to work Monday morning. As the summer passed, we kept working the fields. I thought we were doing a pretty good job. There was always lots more to do than we had time for, but we kept the grass down. As always, Brooks and I milked and fed all the animals before we went to the fields. Then we did the afternoon milking and cleaned the barn and the lot around it after we knocked off from the plowing or hoeing. After that, we showered in the back yard, ate and hit the sack, getting ready for another day's work. All the while, I was making my plans for a visit to Okena, but I kept my mouth shut. I just didn't mention it again.

On the last Friday evening in August, Daddy started giving us his weekly orders. "Next week, I want you to start cutting hay. I want you and Brooks to take sickles and cut the corn stalks and grass in the far field. When you get done cutting there, move over to the next field and start cutting the alfalfa. Cut until it's all clear. Let it dry for a couple of days and then begin raking it into piles...."

I interrupted with, "But, you remember, Daddy, you promised me I could go to Okena to visit Aunt Kate. This is the last week before school starts again and that's what we agreed."

"What do you mean? There's hay to be cut and raked, and if it's not stored in the barn, we'll lose it."

"But we talked about it at the beginning of the summer and you promised."

I was really disappointed.

"That don't matter. That hay's got to be cut and raked. If it's not stored in the barn, it'll be ruined." Daddy was hell-bent on getting everything he could out of the farms.

My mind started doing somersaults. I knew I was expected to take care of the farm, and I knew that my part of the work was important. I wanted to do my share of the work, but then I had dreamed all summer long about visiting Aunt Kate and my cousins.

Suddenly, I thought of a way to make everybody happy. I had always wanted to see the orange groves. I Especially wanted to see them when the oranges were getting ripe. The winters were too cold for any citrus trees except maybe kumquats to grow in Ashburn. Daddy had several of those trees on the farms, but those were the only citrus trees I had ever seen. Still, the thought of seeing bushes loaded with ripening oranges did something for me. I wanted to see the big juicy fruit hanging from the branches when the rinds were just beginning to change color. They would be just right for picking around Christmas time. Dreaming of trees with fully grown oranges, I blurted out my new plan.

"What about if I put off the visit to Aunt Kate until Christmas? I could go during the holidays."

"Gee, that's a great idea! That way, you can cut the hay and get it in the barn next week. Sure, that's aw'right."

So it was back to the fields. Brooks and I cut and

stacked the corn stalks, and we cut most of the alfalfa. All the hay we were able to cut was left in the fields to dry for a couple of days. We then raked it into piles, loaded it onto wagons and carted it to the barn. There was still plenty of grass in the fields when we finished. The next week, Preacher and Virgil plowed the fields and planted a late crop of corn while Brooks and I went back to school. Christmas was three months away, so for now, it was back to our routine. Daddy and I never talked about the vacation again, but I was still making plans. There was still plenty of hay in the fields, and more corn and hay were growing.

Just after supper on Friday, the day Christmas holidays were beginning, Daddy started barking out orders for the next week's work. We were sitting at the supper table.

"On Monday, you need to get out the sickles and start cutting the rest of the hay in the field over beyond the side of the house…."

"But Daddy, I'm going to Okena. Remember? You promised in August."

"Don't matter. You can't go now. There's too much hay to be gotten in."

My mouth dropped open, but nothing came out. Not a word. This was the second promise Daddy had broken. I was floored. My image of him was shattered.

I left the table and climbed the stairs to the bedroom I shared with my brothers, all the time with my head bent down. When I got to the bedroom, I dug out a tattered old satchel and filled it with socks, underwear, a comb, a toothbrush, two clean slipover shirts and some worn khaki pants. I shoved the satchel under the bed and crawled under the cover, pulling a sheet up so that nobody could tell that I still had my clothes on. I said "Hi!" to my brothers when they came in, but I kept the sheets pulled up.

After everybody was in bed and all the lights were out, I listened until I could hear snoring from the bedroom across the hall where Mama and Daddy slept. Then I grabbed that satchel and crept down the stairs and out the back door. It was bitterly cold. God, it was cold, but I was determined. I was going to have my vacation.

Carrying my grip, I walked the two miles to the filling station on U.S. Highway 41, the highway that ran from South Georgia into Central Florida. In the back of the station, sitting on orange crates or kegs of nails, were the man who owned the place and five local farmers all huddled around a potbellied wood stove. There was a brass spittoon sitting in front of the stove. Three of the men were chewing plugs of Bull Durham. Every few seconds, one of the men would push his tongue between his teeth and press it against his lips, suddenly drawing it back, squirting out a stream of tobacco juice aimed toward the spittoon. They almost always hit the target with a few drops of brownish nicotine-laden spit. On average, at least as much tobacco juice peppered the wooden floor around the spittoon as reached its mark. I sauntered up behind one of the men and announced my presence with a simple "Hi."

"Hi! Ain't it a little late for you to be out?" called out Marty Jones, one of the farmers sitting to the right of where I stood facing the stove.

Rocky Ballard, a peanut grower, seeing my satchel, called out, "Where're you going?"

"I'm gonna take a little vacation over the holidays. I'm going down to Okena, Florida, to visit my aunt."

"I think you're a little late. The last bus has already left."

"Oh, that's aw'right. I'm gonna try to hitch a ride."

"Well, draw yourself up a crate and set a spell. Somebody'll be coming along purty soon."

It wasn't more than ten minutes before a flatbed truck with slatted side panels drove up. The driver and his son hopped out. They were Tim and John Young. I recognized both. John, the son, grabbed the gas hose and stuck the nozzle in the tank, then started working the pump handle back and forth. His daddy raised the hood of the truck and held a lantern up until he found the oil stick. He wiped it off with an old rag he found by the gas pump and put it back in the crankcase, only to pull it out again and check it before he drawled, "We got enough oil, at least to get us down there."

After he put the stick back and set the lantern on the ground near the back of the truck, he ambled inside the station and joined everybody else around the stove. That's when I saw my chance. I caught Tim's attention by grabbing his sleeve and asked, "Where're you going, Mr. Young?"

"We're going to Tarpon Springs, Florida, to get a load of fresh mullet."

"You wouldn't be going by Okena, would you?"

"Yeah. We can."

"Can I hitch a ride?"

"Well, we'd be glad to have you, but I have to warn you. There's no room in the cab with John and me. You're welcome to ride in the back. There's some croaker sacks and a tarp you can slide under to keep out of the wind, but it's cold, and we've been hauling fish in there and it smells kinda bad."

"Don't matter. I'd sure like to ride."

As soon as John finished pumping gasoline, he ran in the station and told his dad how many gallons he pumped in and how much to pay. He only had time to pass his hands over the stove one time before we were running out the door.

I crawled up on the bed of the truck and pulled the tarp over me. I slid as close to the cab as I could. The old man was right. The fishy odor was awful, once I got under the cover, but it was cold, freezing, especially when the truck started down the highway and the wind came blowing across the back. Still, I was determined. This was a ride, no matter how uncomfortable it was or how bad the odor. I was staying with it. I bedded down between the layers of tarpaulin and pulled up some empty croaker sacks around me. There I was for the night. I could stand most anything for a ride to Okena. Pretty soon, I got used to the odor and it wasn't so bad. It was still cold, as long as it was dark, but it wasn't quite so bad after daylight. In fact, by the time we reached Okena, the temperature must have been in the fifties. Okena is in Central Florida, about two hundred-fifty miles south of Ashburn. The afternoon sun and Okena's location, halfway down the state, combined to make the temperature almost cozy.

It was nearly two o'clock Saturday afternoon when the truck hobbled to a stop on the Okena town square. I crawled out from under the tarpaulin and jumped down on the street, still gripping my satchel. After saying my "Thank yous," I was left standing there as the truck lurched back in the direction of the highway. There I was, standing all alone in the center of a town I had never even seen before. Still, I had faith that my Aunt Kate would come to my rescue, that is, if I could get in touch with her. My best hope was to call her on the telephone.

As I looked around the square, I saw a phone booth on the corner next to a grocery store. I pulled the handle to open the booth and slid inside. What a relief. At least I was out of the wind. Once inside, I was facing a black metal box with the telephone mouthpiece mounted in front. The receiver was hanging on the right side of the box, and there was a crank just above the arm which held the receiver. I found a slot in the upper right side of the box, big enough for nickels, dimes and quarters. I searched all over the booth but could not find a phone book, so in desperation, I picked up the receiver, held it to my ear and turned the crank. The operator asked, "Number, please."

I had to admit ignorance. I stammered, "Operator, I'm sorry but I don't know the number. I'm in a phone booth on the town square and there is no phone book here."

The operator was so nice. She asked, "Who are you trying to call? Maybe I can help."

"I'm trying to call Mrs. Kate Perry. She's my aunt and runs Mrs. Perry's Sandwich Shop."

"Oh, that's easy. I'll ring her for you." Then she added, "I think your aunt's at home. I just called her a few minutes ago. I'll get her on the line right away."

Telephone operators, in those days, knew everything about everybody. They were really friendly and always helpful.

The phone rang only once before Aunt Kate answered, "Mrs. Perry's Sandwiches. Can I help you?"

It was now the operator's turn. "Please deposit a nickel for three minutes."

Again, I was grappling. I turned my pockets inside-out before I finally found a nickel which I dropped through the slot.

"Clink" was the noise it made as it fell in the metal box.

With that, the operator announced, "Your aunt is on the line."

I took in a deep breath and blurted out, "Aunt Kate, this is Newt Sexton from Ashburn. I came over here to Okena for Christmas. I'm down at the town square, in front of Morton's Grocery. Can you tell me how to get to your house?"

"Newt! It's so good to hear you, but never you mind about the way. You just stay right where you are. I'm coming down to pick you up. You might want to step inside the grocery store to get out of the wind. If you stay near the front door, I can see you and will honk."

I did go inside the front door of the store but stood right there and watched for Aunt Kate. She must have dropped everything because it wasn't ten minutes before she drove up in her almost new Model A Ford coupe with a rumble seat in back. Boy, was I glad to see her! She seemed just as glad to see me, but I didn't see how she could possibly be as happy as I was.

I opened the door on the passenger's side and climbed in. I hardly got the door closed when Aunt Kate wrinkled her nose, then sniffed and wrinkled her nose again as she asked, "Where in the world have you been?"

"I been riding in the back of a fish truck. God, it was cold! I had to lie between some croaker sacks with one tarpaulin under me and another on top."

"I can believe that. You smell just like rotten fish!"

It took us just over five minutes to get to Aunt Kate's house. When she drove up, she wasted no time in ordering me around, "Follow me. I'll show you where the bathroom is. I'll get some fresh water for you. Get out of those clothes while I get a kettle of water and a towel and wash cloth and a bar of soap. Let me have your clothes. I'll wash them while you are taking a bath."

I set the satchel on the floor and opened it to get out some clean underwear and the only other pants and shirt I had. After Aunt Kate poured the water in the wash tub, I had to get another kettle of cold water from the kitchen to cool what she had poured in. After I dropped my clothes outside the door, she grabbed them and was gone.

By the time I had my bath and put on some clean clothes, she was back. She put her arm around my neck and said, "Let me show you the house. We just finished painting the inside a couple of weeks ago."

By that time, my nose was sensitive enough to pick up the odor of the fresh paint.

Aunt Kate then took me on a tour through her four-bedroom home. The best part was when we got to the room on the side where they made "Mrs. Perry's Sandwiches." Aunt Kate prepared and sold sandwiches wholesale. Her business was fast before the Christmas holidays, would slow between Christmas and New Year's, but would be back in full swing after New Year's.

It was while we were walking through the house and into the sandwich shop that I stuttered as I said, "Aunt Kate, I got to tell you something. Mama and Daddy don't know that I'm here. I was promised a vacation, first in the fall just before the end of the summer, but Daddy decided that there was too much hay to be gotten in then, so I said I'd put it off until Christmas. Daddy agreed and I made my plans."

It was just then that we got to the sandwich shop and Aunt Kate stopped in front of the ice box. She had some of her finished sandwiches on a wire shelf inside. She opened the door and pulled out a couple of them and handed them to me, "Here, taste these. They are some sandwiches left over from this morning. You must be hungry, so you can see what they taste like."

She couldn't have been more right. I was hungry as a bear. I took both sandwiches but laid one on a table in the center of the room when she handed me a pint of milk. I bit into the first sandwich and gulped down half the bottle of milk. The sandwich was filled with chicken salad. Boy, was it good! Super good! I finished off that sandwich with just one more bite and grabbed the other sandwich which was filled with pimento cheese, and swallowed it before we finished my tour. It was just as good as the first, maybe better. Maybe my judgment was clouded, a bit, by having had nothing to eat in nearly 24 hours.

In the meantime, I was able to tell Aunt Kate more about my leaving home. "Yesterday evening, when I got home from school, Daddy started telling me about more hay he wanted me to get in. I reminded him that he had promised me a vacation so I could visit you over the holidays. I reminded him that I wanted to see the orange trees when the fruit was getting ripe and this was the time. He got upset and told me that I had to stay and work, since there was more hay to be gotten in. Mama agreed. I was so upset by the broken promises that I decided to come anyway.

"I went up to my room and packed my clothes. Then after they went to sleep last night, I sneaked out of the house and hitched a ride down here. I wanted you to know about that. I sure don't want to cause you any trouble."

"Don't worry, Newt. I know your folks are kinda hard

on you. You're welcome to stay over the holidays, or as long as you want for that matter. Don't worry. Have a good time. I imagine we'll hear from them."

That Christmas vacation was a blast! I had a good time getting to know Aunt Kate better and getting to know my cousins, whom I had seen only once or twice, and then only during brief visits when they had come up to Ashburn. I had the time of my life. Everybody seemed to have a cheerful attitude. So different from home. Christmas day was a big celebration, one that was a lot bigger than we usually had. Both my cousins—Eileen, who was about my same age, and Jeb, who was eight years older than I—and Uncle Gus were there. Jeb was a sophomore in college at the time but home for the holidays. Early Christmas morning, the family exchanged presents. Aunt Kate made me feel right at home. She had some socks and another shirt all wrapped up for me.

The next day, Aunt Kate took my cousins and me for a ride through the orange groves. I was tickled pink to see the orange trees with all that fruit hanging down. We also drove through some other orchards where we saw grapefruit, satsumas, limes and lemons. I'd never seen so much fruit hanging from trees, really looking more like big bushes.

While I was in Okena, I tried to show my appreciation. I had only enough money for my bus ticket back home, so I couldn't buy presents for everybody, but I was able to help clean the sandwich shop and even mix some of the spreads for the sandwiches. That may not have been a lot of help since the work was so slack for the holidays.

"Would you like to go out and meet some of the town people?"

My cousin Jeb was inviting me to walk around the town. We walked down to the square where the fish truck had left me. We walked along the sidewalks in front of the stores. Along the way, Jeb told me about some of the different folks there. He showed me the train depot, so much bigger than the little station in Ashburn where Daddy hopped on the freight each morning. There was a large waiting room with three ticket windows although only two of them were open at the time, and there was only one agent selling tickets. I heard a lady ask for a round trip to Gainesville and then a man behind her ask for a ticket to Orlando. Just then Jeb caught sight of Max Sturdivant, an old friend who was in school at the University of Florida. They acted like two cousins who hadn't seen each other for years.

First, Jeb introduced me. "I want you to meet my cousin Newt Sexton. He's from up in Georgia and visiting here over the holidays."

"Glad to meet you," Max came back. He seemed to be a nice fellow, but just then Jeb interrupted.

"Tell me, what you are studying now?"

"Oh, I'm studying agriculture. I really don't know why I chose that. I've lived on the farm all my life so I thought I might learn a little more about it, what to feed the cows to make them give more milk and what to feed the horses and how to train them. It's a lot of fun, and I think I'm learning a little. Maybe I'll be a better farmer when I get out. What are you studying?" Max asked.

"I'm in Liberal Arts. Just trying to learn more about history and English. I don't know what I'm going to do when I get out. For now, I'm teaching swimming out by Sparklin' Waters. Maybe I'll be able to put on some shows," said Jeb.

Max seemed to be getting a bit fidgety. Then he piped in, "Gee, I've got to get some things at the grocery store and some more at the drug store, so I guess I had better get rolling."

With that, he turned to walk away, but then turned back when I said, "Gee, it's been good to meet you. It's super nice to meet one of Jeb's friends."

Jeb piped in, "It's been good to see you. We'll probably run into each other again before we go back to school."

Max yelled back over his shoulder, "It has been nice. I'll be looking to see more of you later, Jeb."

After Max went inside the grocery, I blurted out, "Jeb, I'd sure like to see what Sparklin' Waters looks like."

"Oh, I'd like to show it to you. Come on, let's walk home and get some bicycles and ride over there so we can look around."

When we got home, Jeb got his bicycle and I borrowed one from Eileen. We rode over to the resort to see it close up, including the glass bottom boats plowing down the river. We didn't ride any of them, but Jeb told me how one could sit around the center of the boat and see all the fish and underwater plants, and even the differences in the depth of the river underneath the boat.

After a week, it was time to go back to Ashburn. Late Saturday night after Christmas, Aunt Kate took me down to the bus station where I bought a ticket and climbed aboard a Blue Lines bus headed north. I had little more than enough money to pay for the ticket. How different the trip home by bus was from my trip down to Okena!

The bus arrived in Ashburn about three o'clock Sunday afternoon. The driver let me off by the old swimming pool on U.S. Highway 41, about a mile from home. After I walked the distance, I opened the back door and found Mama and Daddy sitting at the kitchen table drinking coffee. Before I could say a word, Mama asked, "Well, did you have a good time at your aunt's?"

"Yes, Ma'am. It was really beautiful and super nice."

Those words were barely out of my mouth when Daddy piped up, "That's good,' cause that hay you left in the field is still there, right where you left it. The corn stalks and alfalfa have got to be cut and the hay raked, and it's all got to be stored in the barn. You can start to work on it first thing in the morning."

"You mean I can't go back to school?"

"Not 'til you get that hay in the barn."

I felt really cheated. Here I was, just getting back from a vacation I was promised, and I was being told that I'd have to stay out of school to do makeup work on the farm. I didn't say a word. I just took my satchel up the stairs to my room and set it on the hearth in front of the fireplace. I didn't even open it. I talked to my sisters and brothers when they came in and pulled the covers over me when it was time to go to bed. I was still wearing the clothes I had on when I left Okena.

About eleven o'clock that night, I retraced my steps of the week before. I hit the road back to Okena, this time for good. When I got to the gas station, I found the same old fish truck which I rode the first time standing by the pump. The Youngs were on their way back to Tarpon Springs for more mullet. When I asked for another ride, they agreed but told me that this time they were going directly to Tarpon Strings but could let me off at a crossroad that went over to Okena. That was fine with me. After we got started, I asked whether I might ride with them until they got to Tarpon Springs and then get off on the way back. I wanted to see what the fishing docks looked like, and I wanted to watch the fishermen load the mullet. Going down to Tarpon Springs, we stayed on Highway 41 until we reached Brooksville, Fla., where we cut across to U.S. Highway 19 by a state road. Highway 19 took us directly to Tarpon Springs. We got there about two o'clock the day after New Year's.

I looked the place over. The corrugated tin building was at the end of a long pier where two fishing boats were docked. There was a big vat filled with ice and fresh fish just inside the sliding door of the warehouse. Two fishermen began dipping into the vat and loading fish onto one of the tarpaulins on the back of the truck. When they finished, they pulled the edges of the tarpaulin up and ran a cord through the holes in the lining and tied it all together. There was not much room for me to sit on the way back, but I was able to clear out enough space to make do, at least for the hour I had to ride.

While the fish were being loaded, a cook was frying some fresh fish in the back room. He even baked a pan of cornbread and put on a fresh pot of coffee. We gobbled down every bit of it. We ate just like we were dying of hunger, which I thought I was. I didn't drink the coffee but gulped down two glasses of water. By that time, the fish were loaded and we were ready to head north.

The Youngs stopped to let me off at Juliet, where there was an intersection with a state road going over to Okena. After I jumped off the truck, it backfired twice as it bolted up the highway on its way back toward Ashburn.

I looked around the highway junction and discovered a railway trestle a few hundred feet down the road leading to Okena. Since I was shivering from the cold, blustery wind, I trotted down to the underpass and found a spot that was shielded from the wind. There it was between the mounds of dirt at the base of the railroad trestle. I looked around and found some dry twigs to build a fire. Just after I collected them in a pile and had a blaze going, a car stopped and the driver leaned out and yelled.

"Where're you going?."

"To Okena."

"Get in."

After going the twenty-five miles to Okena, the driver pulled up in front of Aunt Kate's house and let me out. By then it was almost dark.

Just as I was getting down from the car, I turned to the driver and asked, "How much do I owe you?"

"Nothing. I was happy to have your company."

"Gee, I thank you over and over a million times."

I walked up to the front door and knocked. When Aunt Kate opened the door, I announced, "I'm back. This time, I have come to stay."

As soon as I passed her going into the front room, she ordered, "Get out of those stinking clothes and take an all over bath. You've got to dress in some clean clothes."

When I came out of the bathroom, we ambled out to the sandwich shop.

Since I had come to stay, I went right to work. Aunt Kate made twelve kinds of big, fat, juicy sandwiches, and there was plenty of work for everybody. Uncle Gus began work around midnight. Aunt Kate followed an hour later. She woke me up on her way out to the shop. Within minutes, I was wrapping, banding and packing sandwiches for out-of-town delivery. Cousin Eileen also wrapped. By this time, Jeb had gone back to college.

During the next week, I learned a lot about sandwich-making. We packaged between twelve hundred and two thousand fresh sandwiches every day. Bread for the sandwiches was baked in Spartanburg, S.C., from a special recipe. It had an unusual texture, slightly coarse but still moist enough to stay fresh. It arrived at the depot by Railway Express every morning just before daylight. There were way over a hundred loaves, six-by-six by twenty-four inches. I drove Uncle Gus to work at a canning factory at six forty-five on weekday mornings. On my way back, I made deliveries to whatever snack shops and filling stations were open at that time of day. I picked up the bread at the train station, and when I got back, I sliced every loaf, trimmed off the crust, and repackaged it, just as it came. I saved the crusts and added small bits of leftover fillings. Were those bits of sandwiches ever good! Ideal for a growing boy.

After I finished trimming the bread, I loaded boxes of sandwiches for deliveries out of town, for Gainesville, Inverness, Dunnellon, Brooksville, Tampa, Saint Petersburg and Clearwater. I drove them down to the bus station. Afterward, I drove through the alleys around town looking for clean boxes in which to pack more sandwiches. I carried them back to Aunt Kate's and helped her with her household chores and yard work.

It was about the middle of February or the first of March when Aunt Kate suggested that I ought to think about going back to school. So one morning after I delivered the sandwiches to the bus station, I went by the school and registered. I was put in the sixth grade class and began going regularly, but after that I was not able to get around to the places to collect the boxes for packing sandwiches in the mornings. I tried to do that in the afternoons after I got out of school, but by then, many of the boxes were gone and many were broken down, so

Aunt Kate had to go collect the boxes in the mornings. I was not able to be as much help to Aunt Kate and Uncle Gus as I had been.

One Saturday morning after the sandwiches were delivered to the bus station and I went back home, Uncle Gus called me outside and told me, "We've got to clean out the septic tank."

He handed me two five-gallon buckets, and he carried two others and a rope. He opened the top of the tank and began dipping out the honey. He told me, "I'll dip out the tank. You carry the buckets down yonder in the woods. Be sure it's far enough down there, so it won't stink up the place."

What a mess! Uncle Gus dipped out the stuff and I carried the buckets into the thicket. It seemed like a mile, but it was probably only about 500 feet from the house. What a stinking job!

I was happy in my new home, and I thought the family was pleased with me until one Saturday morning in late March or early April when Uncle Gus, just out of the blue, stunned me. He said, "I believe your visit has been long enough. You need to find another place to stay."

2

UNCLE GUS NEVER explained his urging me—no, telling me—to leave my newly adopted home, and I could never bring myself to ask for a reason. I just never understood why I was suddenly being pushed out the door. I wondered whether my marching orders were given because I went back to school and was not able to help with the business as much as before. I wondered whether Aunt Kate had received word from my folks back in Ashburn and if they were trying to get me to come back home. No matter what the reason, I didn't want to stay in a place where I was not wanted. I had to find another place to live and I had to find a job. I was not going back home.

    I really didn't know what to do. I grabbed my satchel with the few things I had and wandered up and down the streets of Okena. As I was walking around the town square, the All American Grocery store caught my eye. It was a retail and wholesale business. I went in and asked the manager for a job. I was lucky. It was Saturday and he needed somebody that day to help sack groceries. I was hired on the spot. Mine was a part-time job. I was to work only on Saturdays. I began sacking groceries at seven o'clock in the morning. I worked at the check-out counter until about three in the afternoon and then began stocking shelves until they were full. Sometimes, my work was not finished until three o'clock Sunday morning. I was paid three dollars for eighteen hours of hard labor.

FINDING A NEW place to live seemed like it might not be possible for a fifteen-year-old kid with no family. For a few days, I rented a room in a boarding house, but that took almost all the money I was making, so I knew I couldn't stay.

My cousin Jeb, who was at Forman College, had a canoe. Fortunately, he came home on weekends, and he was good enough to lend me his canoe. He even showed me how to pole it. I practiced pushing it up and down the river. Poling a canoe takes a lot of practice since canoes will tip over very easily, but with practice, I got the hang of it. It was about two years later that my poling won me the respect of an Indian chief. That was after I was working at Sparklin' Waters.

As with a lot of my other shenanigans, I hit it lucky. On my second trip up river, I ran across a little broken-down jetty sticking out over the water about a half mile from the headwaters of the river. A hundred yards from the edge of the river was an old unpainted, two-room frame shack sitting on broken brick pillars. It was obviously abandoned. The door to the screened front porch was standing open. Many of the screens on the porch were torn, the upper parts hanging free. Several of the screens around the porch, as well as those on windows on the side of the house, were covered by cloth, and even bits of the cloth were torn into shreds. Also many of the window panes around the house were either cracked, broken out or just gone. The grounds around the house were littered with trash: old rags, newspapers, and bottles, even parts of broken-down chairs and benches.

As if that were not enough, there were bricks and even piles of gravel and rotting tree limbs lying around.

I tied the canoe to the jetty and walked up to get a closer look. The house and the yard around it looked awful, but the house had four walls and it had a roof. I could see holes in the tin roof over the porch, but I couldn't see through the roof over the inside. Besides the front porch, the house had two rooms, one in front and the other behind. A single brick chimney stood in the middle of the wall separating the two rooms. There was a fireplace in the back room but only an inlet for a flue in the front room. The only furniture in the shack was an old wood-stoked, potbellied stove sitting near the back wall of the front room. A rusty flue pipe connected it to the chimney that vented the fireplace in the back room. I figured I could use the stove to cook meals if I ever got enough money to buy food. When wintertime came, I might need it to heat the house.

What more could a homeless kid ask for. Clearly, no one lived in that house. It must have been vacant for a long time, maybe months, maybe years. Right off, I knew that was going to be my new home. I had never known any squatters, but I was about to be one.

Inside the house, I realized that someone must have built a fire in the center of the front room. There was a hole the size of a number two wash tub burned through the floor. But overall the house was priceless. Cold weather was not a problem, at least for the next seven or eight months, This was going to be my home.

There was an outhouse about thirty-five feet behind the shack and to the right as I faced the house from the jetty. The outhouse was no more attractive than the main eyesore, but it would serve its purpose.

Sedge grass was knee-high all over the yard, covering at least that portion of the ground not buried under litter. In fact, the sedge grass hid most of the litter. I soon trampled down the weeds between the shack and the outhouse and between the shack and the pier. I collected the rags, newspapers and tree limbs and burned them. Then I collected the glass bottles and tin cans and whatever other rubbish that could not be burned and stacked it all in a neat pile behind the house. Later, I was able to bring home a Laddie Boy swing blade to cut the weeds and briars. To me, it looked pretty nice but even months later when my boss saw it, he dubbed it "Camp Nasty," a name by which it was known as long as I lived there.

The day after I found the shack, I moved out of the boarding house and set about making this my new home. I scrounged a few boards and some nails and I borrowed a hammer and hand saw from Jeb and built a bunk, which I hung on a side wall in the front room. I even got a needle and some thread and sewed a pad for the rack using croaker sacks which I filled with straw. With all this, I was able to make a fairly comfortable bed. I even shaped a pillow, using more of the same bags stuffed with straw. Later, I built another bunk bed in the back room where my cousin Eileen spent nights when she stayed out late with her boyfriends.

It was only a week after I moved into the shack that a polecat climbed in through the hole in the floor and left his calling card. I was asleep at the time, but when I woke up, what a foul odor! It took me only a few minutes to grab the skunk with an old rag and toss him into the river. My next chore was to find some old planks to cover the hole. That kept out any other varmints, at least the size of weasels.

    Camp Nasty was the old home of the famous Aunt Lila, a black psychic who had died five years before I came upon her abandoned shack. She claimed to be 109 years old the year she died. After her passing, all the furniture in the house except the stove was stolen and the house was deserted except for passersby who threw rubbish in the yard.

    For seven years that shack was my home. Its major attraction other than the price, the walls and a roof was a scenic view of the Sparklin' Waters River.

3

"**MAWNIN', CAPTAIN MAYSON.** I'm Newt Sexton, an' I'm looking for a job. I want to work here at Sparklin' Waters."

"God! You look mighty young. How old are you?"

"Fifteen. I'll be 16 in August."

"Gee, you're even younger than I thought….Naw. You're too young to work here. This work's too hard for a kid your age."

I was floored.

I had memorized my lines the night before, hoping I'd make a good impression and maybe even get an agreeable answer when I visited, but his rejection was so quick and sounded so final that it left me dumbfounded. I didn't know what else to say. The time was eight-thirty on the third Tuesday in April.

That was my first time meeting with Shorty Mayson. He was the man who did the hiring and firing at Sparklin' Waters, was a resort built a couple of miles outside of Okena, in the middle of Florida. It surrounded the headwaters of the river by the same name. Shorty's office was on the second floor over an antique shop in an old frame building which had recently been renovated. The building also housed rooms for swimmers or sunbathers to take showers and change clothes before and after swims or after just lying in the sun on the beach. The outside of the building had a nice stucco finish making it look like the two other buildings beside it on the waterfront. A coffee shop was in the building next door, but I avoided that place. I didn't dare go in any place where I might be expected to part with even a penny.

    I needed money, enough money to buy whatever would stave off my hunger. I hoped to find a job near my new home. That morning, I was intent on finding such employment.

    Sitting there in front of the captain, I looked around. His walls were lined with bookcases in which books and ledgers were stacked in a haphazard fashion as though they had just been thrown at the shelves. Titles of those I could see from where I was sitting seemed to concern the operations of companies or parks that catered to tourists. I saw one book on underground water channels, but the words were so long, I didn't get the hang of what they were about. The shelves needed dusting, but most of all, the books needed to be straightened up and put in order. A couple of books and another ledger were lying on the captain's desk along with some pencils and quills and even an inkwell. Newspapers were scattered on the floor around a trash can next to his desk. In short, the office looked a mess. It needed somebody to straighten it out. Those were things I could do, if only I could get the captain to hire me.

    He was a middle-aged, five-foot, eight-inch chubby man, and he held a stubby pipe between his teeth. He sucked in on the pipe and then blew out a big puff of smoke from time to time even during the few minutes I was there.

    I just knew that there had to be something I could do at Sparklin' Waters. Even though I was a teenager, I had worked for years, at least for the last five years. While I was in grammar school back in Ashburn, I had done more than my share of chores on the farms my daddy owned. In earlier years, I had helped with the housework, but during the last five years, I had milked cows, cleaned barns, cut weeds, plowed fields, and cut, raked, and stored the hay in a barn. I drove Daddy's car from the time I was 11 years old, and when Daddy was away from home, I passed his orders on to the hired men. Sometimes I had to change his orders to keep the work running smoothly. After I left home and came to Okena, living at Aunt Kate's, I helped make and deliver sandwiches. I took the sandwiches wrapped individually to the shops and filling stations in town and took those packaged in cartons to the bus station for delivery in surrounding towns. I drove the family car to take my aunt's husband to work and to deliver the sandwiches. I even toted the honey dipped from the cesspool into the woods behind the house while Uncle Gus was dipping out more. I despised the job, but I did it, proving, at least to myself, that I could do most any job I had to do.

    I tried to think of something else to say, something that might convince the captain that I could, at least, help with work at Sparklin' Waters, but all that came out of my mouth was "Ah, ah." When I realized that there was nothing else to say, I got up and shuffled out the door and down the stairs onto the lawn in front. I didn't know where I was going. I was just leaving his office, but I knew I'd be back.

    When I got outside, I decided to take a look around the place. The park was huge near the place where clear water from a gigantic underground aquifer gushed out of a cavern forty feet below the surface. The cavern could be seen only by looking straight down from the water's surface. That meant, to see it, I'd have to be in a boat. People could see the fish and other beauties of the river bottom from open boats floating on the river, but special boats were needed to show it to large numbers of people, and it took a lot of paying customers to make the viewing pay off. The Sparklin' Waters resort solved this by developing a fleet of boats with glass bottoms powered with gasoline engines. Passengers could sit around glass windows in the floor of the boats and see the fish and wonders of the floor of the river as the boats chugged back and forth. To keep out competition, the owners of the park didn't allow any other boats on the river when the glass-bottom boats were plowing the waters.

That morning, as I looked around the grounds, I saw the three buildings lining the headwaters. They were only a little over one hundred yards from the entrance to the resort from U.S. Highway 40. A driveway made out of bricks ran from the gate to the front of the ticket office. The driveway was wide enough for two lanes of cars. The lanes separated and curved around a flower bed, where six rows of pansies bloomed year-round. The little flowering plants were growing around a bank of palmetto palms. Among the pansies were some other plants with white blooms and even some with multicolored blossoms. I didn't know what they were. I later came to realize that the pansies were rotated so frequently that they gave a lively appearance all the time.

Cars could park anywhere along the roadway, but there was a parking lot at the end of a connecting driveway about one hundred fifty yards beyond the circle. It was south of the row of buildings and behind the sand beach in front of the swimming area. Looking down the river, there were cypress trees lining the water on both sides, giving a spectacular view in the direction of the shack I had taken over. The trees seemed to stretch skyward from the swampy surroundings, which were covered with thick underbrush and reedy grasses. I just couldn't name all the green bushes that lined the river's edges.

At the front of the water's edge along the driveway, the first building nearest the entrance was the combination ticket office and souvenir shop. Next was the coffee shop-restaurant and third was the antique shop with shower stalls in the back. The ticket office-souvenir shop and the restaurant were single-story buildings, while the antique shop was on the ground floor of a two-story building, with the offices of Captain Mayson and Mr. Mack Tracy on the second floor. Each of the buildings had a stucco finish. They looked like they had been imported from Mexico. Three windows were on the front of the ticket office. They took up about twenty feet of the front wall. To the right of the ticket office as one faced the building was an entrance to the gift shop.

Between the ticket office and the highway and extending back from the driveway between the headwaters of the river and the highway was a tropical wonderland, a garden with walkways winding through acres of azaleas, camellias and all sorts of flowering shrubs. There were even a few tropical lilies planted among the shrubs. In the center of this paradise was a spectacular rock garden. The flowers were arranged so that some were in bloom year-round. Behind the ticket office extending over the headwaters of the river were the covered docks where passengers boarded the glass-bottom boats. After buying their tickets, guests could stroll through or around the souvenir shop and onto the jetty where the boats were docked. The pier was big enough for four boats to be docked at the same time.

On the other side of the lake from the boat dock, beginning behind the antique shop and circling around the edge of the water, was the sand beach. A section of water in front of the sand beach, shallow enough to be safe for amateur swimmers, was marked off by a rope tied to cypress posts. Multicolored streamers dangled from the rope. At the back of the swimming area, some twenty yards offshore, stood a tower for the lifeguard. Its platform was ten feet above the water.

Beyond the sand beach and next to the parking lot was an ugly little building that looked like it was about to collapse. It had been painted white, but the paint was peeling, and what was left was so off-color that it was a smudged gray. It got my attention even when I was standing a hundred yards away. It seemed so out-of-place in the midst of such beautiful surroundings. I later learned that it was an abandoned shooting gallery, a place where visitors to the park could polish their shooting skills with air rifles or darts, aiming at bull's-eyes while they were waiting to board the glass-bottom boats. It had become termite ridden and was sadly in need of painting. It was an eyesore sitting right there in the heart of Eden.

Beyond the parking lot were two outdoor grills and some roughly hewn picnic tables where the employees of Sparklin' Waters, supervised by Shorty and his brother, Brad, held cookouts to entertain managers of local hotels, filling station attendants and shop owners, not to mention elected officials from towns and villages around Sparklin' Waters. They might be from as much as fifty miles north, along the highways running south toward Central and South Florida. The cookouts were used to entice businessmen and other townspeople to tell travelers about the resort. It catered mostly to tourists, but the owners liked to attract the locals as well. They wanted to attract anyone who might spend a few nickels, and they didn't object to big money either.

When I looked over this huge playground, I was still thinking about what work I might be able to do. There were grounds to be kept, weeds to be cut, buildings to be cleaned and painted, and the glass-bottom boats to be kept up. I was sure there were hundreds of other jobs that I knew I could do, especially if I had some brief training. I just had to convince the captain that I was needed at Sparklin' Waters. It was true, I was young, very young, and I had not even finished grammar school. I was still better educated than a lot of the laborers who worked the flower beds. Since I had a place to live only half a mile downriver and had no transportation other than my cousin's canoe, this would be the ideal spot for me. I just had to convince Shorty that I was his man.

Still trying to figure out how I could convince Mr. Mayson, I meandered across the beach to the canoe that I had left on the sand at the edge of the swimming area. I pushed the boat back onto the water and climbed in, grappling with the long wooden stick, poling the canoe downstream toward the jetty in front of my shack.

I didn't know that Sparklin' Waters even existed before I visited Aunt Kate at Christmas. Jeb and I had gone out to the place once during my week's vacation, but while I was making deliveries in town, I heard a lot about the history and the development of this natural beauty spot.

The locals bragged that Florida's Sparklin' Waters is

one of the biggest limestone springs in the world. They said that over five hundred million gallons of water flows out of the cavern at the bottom of the lake each day. The water flows at a temperature of seventy-two degrees Fahrenheit. It carries with it hundreds of tons of minerals from the earth below. The amounts were so great that neither I nor any of the people I talked with could make out the numbers, let alone the details. Neither did the locals know from where the numbers came. Still, they were impressive at least to a kid like me. The people just knew that the amounts were gigantic and the temperature was ideal for swimming year-round.

 The big cavern at the bottom of the lake is twelve feet high and sixty-five feet wide. Again, none of the locals I talked with had seen the chamber except from a boat on top of the water, and they didn't know how to take measurements from there. Water flowing out of the cavern formed the head of the Sparklin' Waters River. That was for sure. Overall, the Sparklin' Waters River is nine miles long, and it pours into a branch of the St. Johns River, which eventually empties into the Atlantic Ocean near Jacksonville. The flow of water from the cavern at the head of Sparklin' Waters is enough to keep the river flowing at four miles an hour, and because that water comes from an aquifer, it is crystal clear—at least it was at the time I lived there. Even before I came to live in Okena, motion picture companies found the river is comfortable temperature ideal for making underwater movies.

 The headwaters of the river are about two miles from the center of Okena, which was discovered by the Spanish explorer, Hernando de Sota, in 1539. He was thought to be the first European to ever see Sparklin'

Waters. Since that time, many authors have paid tribute to the beauty of that natural paradise. As early as the beginning of the nineteenth century, people thought of it as a popular tourist attraction.

When Mr. Ed Masters, who was part owner of the Barnum & Bailey circus, was elected commissioner of Indian County, he bought the land around the headwaters of Sparklin' Waters and tried to make the natural setting into a tourist attraction. That was during the early part of the twentieth century. He bought some steamboats with glass bottoms so that passengers could see the plants and animals as well as the beautiful grasses and petrified wood at the bottom of the river, but he was never able to make money from the few visitors to Utopia. It was just five years before I moved to Camp Nasty that Mr. Mack Tracy Sr., who had made a lot of money turning pine sap into turpentine, formed a partnership with his son, Carl, to lease the resort. They were convinced that, with the right people and the right effort, Sparklin' Waters could be made into an attraction that would become known far and wide. Shorty Mayson had earlier turned a small sandwich shop in the railroad station at Okena into a top-notch restaurant. Knowing this, Mr. Tracy walked into his restaurant one day and pulled Shorty aside.

"How would you like to become a partner in a new development here in Central Florida?"

"Well, ah, ah, what do you have in mind?" Shorty wanted to know.

"My son, Carl, and I have just leased the land around the headwaters of Sparklin' Waters River. We are convinced that the right management can turn it into a place that will really appeal to tourists, but we've got to have the right man at the head to call the shots.

We've seen what you have done with this restaurant, and we believe you could do the same thing with Sparklin' Waters. We believe you could make it into a real bonanza."

"What's it worth to you?" asked Shorty.

"Well, we'll give you five years. If you make a success of it, we'll deal you in as a full partner," said Mr. Tracy.

"Sounds good to me, but let me think it over for a day or two. This restaurant is doing so well, that I don't want to just throw it away.

"There's one thing I need to know. Do you have any cash for improving the facilities?" Shorty added.

"Not a lot, but we have some and we have good credit. We're sure Mr. Masters will help us."

Two days later, Mr. Tracy walked into the restaurant. Shorty saw him and, right there in the middle of the restaurant, let out, "Mack, I've been thinking about your offer and I've decided to take you up on it. When can we get together?"

"This afternoon if you are free. Can you come out to Sparklin' Waters? We'll sign you up and give you an office."

"What time?"

"Is two o'clock okay?"

"I'll be there."

It was a done deal. They signed the papers that afternoon. Shorty was given a temporary office in a little shack while the whole place was being redecorated. Permanent offices for the manager and owners were built above a newly decorated antique shop, and from then on, Shorty was manager of Sparklin' Waters, a job he held for the rest of his life.

The two Tracys with their new manager bought new gasoline-powered glass-bottom boats and planted flower gardens in addition to remodeling the building with the antique shop on the ground floor and the offices upstairs. They put in the grills and picnic tables to entertain VIP's and people who might meet tourists around the area. From then on, they put on a spirited public relations campaign. The business turned out to be more successful than anyone had dreamed it could be.

The new glass-bottom boats were twenty-seven to twenty-eight feet long and carried thirty passengers. They were powered by two sixteen-horsepower horsepower gasoline kickers, which were really Johnson outboard motors. In the center of the floor of each boat was a solid plate of glass one inch thick, two feet wide and fifteen to eighteen feet long, sealed around its edges. Riders could sit around the glass plate and watch all the marine life and the grass and petrified wood on the river bottom underneath the boat as it motored up and down the river.

Improvements in the tourist attraction continued through the early 1920s. Soon the natural beauty of the place surrounded by the ornamental construction and newly planted gardens caught the eyes of three investors from New York who submitted a bid to sublease the place. They pictured a gold mine and offered a contract to take over the place for $3 million; $1 million in cash and the rest in payments over five years. Shorty jumped at the chance and deposited the down payment in a local bank. Fortunately or unfortunately, the new investors went bankrupt, and management of Sparklin' Waters went back to the original threesome. The down payment, which Shorty had stashed away, made a nice nest egg, which gave the owners more money to make further improvements in facilities and equipment while the tourist attraction kept operating. That was when I came on the scene and started hounding Shorty to hire me.

MY COUSIN JEB held swimming camps on the banks of the Sparklin' Waters River about a mile downstream from the resort during summer vacations. He held one camp for boys 10 to 12 years old and another for girls of the same age. One camp came right after the other, each lasting two weeks. Jeb's training camps were great. Many of his swimmers became real champs, and he began putting on water ballets at such well-known tourist attractions as Weeki Wachee and Rainbow Springs. Other entertainment centers throughout Florida contracted to have him put on more pageants. I tried to help Jeb with the training camps, doing whatever I could. And whenever I had a free moment, I visited Captain Shorty Mayson. I did that over and over. I begged, "Won't you please give me a job? Just try me."

From the first day, Shorty's answers when I asked him for a job never changed, at least not for six or eight visits. "No, you are just a boy. You're too young. You don't want to work. You just think you do."

I learned some comebacks of my own. "Please give me a chance. I'll prove you wrong. I'll prove to be better than you ever thought possible."

Finally, I wore him down. One day, Shorty jumped up from his desk and tromped over to the window. He pointed to that old dilapidated frame building I had seen near the sand beach surrounding the swimming area and said, "You see that old, run-down house over there? That used to be a shooting gallery where customers could take target practice while they waited to board the glass-bottom boats. That was in the old days, before we had many boats. Now, customers don't have to wait so long, but anyhow, the building is rotting away. We've got to get rid of it. I'll let you tackle that. I want every piece of it torn down and burned. Things that won't burn should be stacked neatly. Let's see what you can do with that."

While he was taking, I ran over behind him and looked over his shoulder. From there, the shooting gallery looked even worse than when I had seen it that first day I visited Shorty. In fact, it was surely the worst mess I had ever seen. It was so termite-ridden that sections were caving in. Tearing it down would certainly improve the appearance of the resort.

I guessed that Shorty thought I could never get the mess cleaned up, but when he gave me the chance, I went right at it. I ripped the building apart, board-by-board, and stacked the rotting boards with all their termites in a pile behind the place where the building stood. Then I set the mess ablaze. What a bonfire! I stoked the timbers until there was nothing but ashes. Within a day and a half, working by myself, I tore the gallery apart and cleared the area of all its rubbish. I gathered up all the unsightly trash that could not be burned and put it in a separate pile to be carted off into the woods. After the grass grew back, the spot became an appealing addition to the park. It seemed to fit right in.

When I walked into his office after I had finished, the captain simply commented, "Looks like you did a good job of getting rid of that old shooting gallery. I like the way you burned the stuff. The place looks good."

That was as big a compliment as I ever got from Captain Shorty, but he carried on, "If you want to keep working here, I'll let you start cutting grass along the river banks. I want you to cut all of it. You should start by cutting the grass around where that shooting gallery was and just keep cutting down the river. Your job will last only as long as there is grass to be cut…. There's one other thing. You are not to socialize with those other laborers." When he said that, he pointed to six or eight men all huddled together in a group beyond the rock and flower garden between the ticket office and the highway. They must have been telling jokes. Every once in a while they would break out in a roar of laughter, but they certainly were not getting much work done.

After those instructions, he took me over to a closet where he pulled out two Laddie Boy sling blades and two files to sharpen them, and he handed them to me.

"Now, you just follow my instructions and we'll see how you do."

At the end of the first week, I got my first pay, four dollars and fifty cents for four and a half ten-hour days. That was the same rate I was paid for months. My days began at seven o'clock in the morning and I got off at five-thirty in the evening. I had 30 minutes off for lunch.

I never understood why Shorty didn't want me to mingle with the other hired hands who were doing most of the manual labor in other parts of the park. I wondered whether they were not getting enough work done when they got together for their yarns. It seemed as though Shorty was taking me under his wing, rather like a son. Maybe he didn't want to dilute my enthusiasm. Even though I didn't understand his reasoning, I followed his instructions to a "T."

Through the summer, I trimmed the reedy grass along the river banks. I was able to make them look more like extensions of the grassy lawns. They blended in, right down into the water. The river's edge, which had looked shaggy, became a lot more attractive. I cut grass every day from the time I arrived for work until I got off.

When the weather cooled at the end of the summer, the grass stopped growing, and I started worrying about my job. When I went in to the office to pick up my pay on the last Friday afternoon in September, Mr. Tracy handed me my money. He looked up with a smile on his face and said, "Newt, you've done a good job, tearing down that old shack and keeping the grass trimmed along the river banks. You have done well. I'm proud of you."

More concerned about where my next paychecks were coming from than any compliments, I answered, "Thank you, Mr. Tracy, for your kind words, but what can I do now that the grass is all cut? Is there any more work for me?" I couldn't bring myself to ask what was really on my mind, "Is this the end?"

"Son, you don't have to worry. We've got plenty of work for you. We'll keep you on as long as you work like you did through the summer. When you come in on Monday, you should report to Mr. Ritchey over in the maintenance shop. He needs some help getting the boats ready for next spring. He's a good man. He'll take you under his wing. If you work well with him, he'll treat you like his own son. Mr. Ritchey's not only good. He's smart. He knows a lot about how things work and he can teach you a lot. Learn all you can from him."

"Thank you, sir. I was hoping you might find something more for me to do."

With that, my immediate worries were over. I went back to my shack and took it easy for the weekend, getting ready for my new job. During that weekend, I thought a lot about Captain Shorty Mayson, the man who had hired me.

I RESPECTED SHORTY as much or more than any other man I ever knew. Shorty was steering Sparklin' Waters into a nationally recognized tourist attraction. Much of the growth was coming about right there in front of me, and I felt like I was a part of it. Shorty ran most of the day-to-day operations, but probably his greatest strength was in public relations.

A native of Waresboro, Ga., Shorty grew up in North Florida and then worked with his brother, Brad, who ran a restaurant in High Springs. After he got a lot of experience running that cafe, Shorty moved to Okena to take over the grille in the Coastline and Seaboard Railway Station, where he started out with a captive flock of customers. The sandwich shop was a big success. He developed it into a first class restaurant.

At Sparklin' Waters, Shorty called himself "The Plunger," but he admitted years later that even he didn't realize the potential of the natural retreat.

Shorty was a real fighter. A real bulldog! But as he grew in fame and became rich, he shared his wealth. For example, when the Okena Salvation Army outgrew its old home, the organization known for its good works built a new location. Their new building, made out of blocks of lime rock mined around Okena, was sixty feet wide and ninety feet long. When the building was complete, a member of the group came out to Sparklin' Waters and asked Shorty, "Could you make a contribution toward the cost of the new building we put up for the Salvation Army?"

"I might be able to. Can you tell me what the building cost?" Shorty asked him.

"Ten thousand dollars."

Shorty never blinked. He walked into his office and made out a check for the full amount and marked it, "Paid in full."

Shorty gave me my first job. At the time, my life depended on finding something that paid me at least a little money. I was just a kid and searching for not only money to eat, but also a father figure. I was disappointed with my own daddy whom I had always previously admired. He failed me when he reneged on not just one, but two promises of a vacation. I was even more upset when he wanted to make me work on the farm rather than go back to school. Shorty took his place. He was a business man, but when someone proved his worth, Shorty stood behind him, all the way.

4

"MR. RITCHEY. I'M Newt—Newt Sexton. Mr. Tracy sent me over here to help you and to learn everything I can from you." I said as I headed—almost running—through the huge corrugated tin building toward the little wiry man leaning over a work bench fixed on the wall of the tool shed in a back corner of the main building. The time was seven forty-five in the morning on the last Monday in September.

Mr. Ritchey was examining a piece of machinery that he was holding in his hands. I later learned that it was a carburetor taken from one of the outboard engines used to drive the glass-bottom boats. At the time, I was so worried about getting to know Mr. Ritchey and working out a good connection with him that I wasn't paying much attention to what he was doing. I was thinking more about making a good impression.

Mr. Ritchey was dressed in his blue denim overalls and matching shirt. His head was covered up by a tattered old panama hat. The sleeves of his shirt were buttoned at the wrist, hiding his arms, like they always did. That was his uniform every day.

Even though Mr. Ritchey's work clothes were neat and clean this early in the morning, his hands were already covered with grease. Little specks of dust coated the grease, especially over his fingertips and in his palms.

Mr. Ritchey barely looked up from his bench when he said, "Yeah, I heard you was coming. Why don't you drag up a box and have a seat?"

His answer was neighborly enough. He was the soft-spoken man who was going to be my boss and teacher. At first, I was puzzled by his lack of get-up-and-go, but I was not going to let any coolness that I just thought might be there stand in the way of good neighborliness. After all, if he was cool toward me, it was probably just because he was so wrapped up in whatever he was trying to fix.

I looked across the room and saw an old empty orange crate laying under a bench out in the main building. I ran over to grab it and got back just as Mr. Ritchey sat up and looked me straight in the eye. He mumbled, "Where y'all from?"

"Well, I was born in Sylvester, Ga., but my folks moved to Ashburn when I was just six months old. I lived there until last Christmas when I came to Okena. I've been living in an old shack I found about half a mile downriver since April."

"How old are you?" asked Mr. Ritchey

"Sixteen last month."

"What brought you to Sparklin' Waters?" he inquired.

"I came to Okena to live with my aunt, but it was kind of crowded." That was a little white lie, but I was trying to steer clear of any answer that might suggest that there was any bad blood. I didn't want to get into why I had left home at Christmas and then was told to find another place to live just three months after I had come to live with my aunt. The fact was, and still is, that I really didn't know why Uncle Gus had told me that I had outlived my welcome.

"How far did you go in school?" Mr. Ritchey asked.

"I finished the fifth grade last year. I was halfway through the sixth grade when I left. I registered for school in Okena and went for two or three weeks but had to stop that when I moved out here to the house down the river and came to work at Sparklin' Waters."

"Why did you quit in Georgia?"

"That was when I moved to Okena to live with my aunt," I answered, hoping that would be all he asked about my schooling, since I was determined not to get into a discussion about the falling out I had with Daddy

"How did you do in school?" Mr. Ritchey asked. I felt sure he was expecting me to say that I flunked out.

"I did aw'right. Passed everything. I liked school."

"What are your plans for now and for the future?"

"I want to work here. I want to learn all I can from you."

The conversation dragged on. It seemed like Mr. Ritchey asked a zillion questions, and I told him everything about my past. I told him that I was the oldest of seven children, born to a railroad conductor who was a real hard driver, always keeping his eye on how much money he could make, or better, how much money he could get out of the work on the farm. I knew for a long time that Daddy was a kind of Scrooge, but I surely wasn't going to say that. I never minded working on the farm as a boy, but I could not excuse my daddy for going back on his promises to me, not just once but twice, and then refusing to let me go back to school until I gathered in all the hay in the fields. That was too much. The memory of Daddy's demands really rubbed me the wrong way, but my mind was made up. I was going to keep my mouth shut about our run-in.

I told Mr. Ritchey that, being the oldest boy in the family, I managed the farms when Daddy was working on the railroad. I mentioned planting the garden and corn, cotton, and peanuts, plowing, milking, cutting and raking hay and hauling the hay to the barn. I explained that I spoke for Daddy when he was away; that is, I told the hired men what to do.

I also told Mr. Ritchey that, after I came to live in Okena, I helped make sandwiches for Aunt Kate and then delivered them to the local shops and bus station. I even told about my part in emptying the cesspool, carrying the honey to the woods, but I only mentioned that. I hoped I wouldn't get another job like that at Sparklin' Waters, but I knew if I were told that I had to do it, I would. I had to eat.

When I told so much about my past, Mr. Ritchey started to loosen up. He told me some things about his past, not in as much detail as I had, but enough for me to begin to feel like I knew the man who was going to be my teacher.

Mr. Ritchey was a Southerner. He had grown up in Central Florida. His upbringing was a lot like mine, only he was older and a lot more educated than I was. He had graduated from high school but was tired of book learning and hitchhiked to Texas when he heard that there were plenty of jobs in the oil fields. He was hired by the Standard Oil Company, first as a gofer or roughneck, but when his supervisor saw that he was good at drawing and overhauling things that were broken down, he was put in charge of maintenance, first for one and later for five crews. He worked in the oil fields for several years. He became a jack of all trades. In the meantime, he married his high school sweetheart, a girl from Wildwood, Fla., and when she got pregnant, she begged to come back to Florida to live near her parents. Mr. Ritchey gave up his job, and they moved to Okena. A few days after he got back, he went in to see Captain Shorty, who hired him on the spot. He was put in charge of keeping up the glass-bottom boats as well as all the other equipment they had at Sparklin' Waters. That was a big job.

Before long, I got to know Mr. Ritchey as an excellent carpenter, painter, plumber and electrician. He was a man who could take care of almost any kind of woodwork around Sparklin' Waters, and he was a real pro when it came to mechanical repairs. Most of all, he had a lot of good common sense. He was a quiet man, thoughtful. Even during our first conversation, he took on a sort of fatherly interest in me. Toward the end of that talk, he said, "If you absorb all the teaching I give you, you'll not have to worry about finding a job for the rest of your life."

I needed that.

I thanked Mr. Ritchey and promised him that I would never give him any back talk. I was not only willing but anxious to do any job he wanted me to do. During the next two and a half years, the tutor and the student had school every day. I am still thankful for this wonderful teacher and lifesaver whose training I find myself still using even now, many years later.

While we were talking, I was glancing around the shop. The walls of the shop-boat service building were made of corrugated steel. The floor of the shop was concrete. The walls were lined by wooden benches about waist high. There was also a bench around the entire tool shed, which was a closed-off section of the service building. A double door opened into the main building.

Lighting inside the tool shed was poor in the center but lamps hung from rafters over the work benches. The rafters crossed the width of the shed about ten feet above the floor. The shaded lights were just three or four feet above the benches. Neatly arranged on the back wall were all kinds of hammers, chisels, saws, screw drivers and wrenches, and there were buckets and buckets of nails, of all sizes, and screws, bolts, washers and nearly every other kind of mechanic's and carpentry tools or material I could imagine. I had never seen most of them. Each wrench and nail and screw had its place so that Mr. Ritchey knew exactly where he could find it.

The tool shed was neat, but the floor out in the main building was a mess. There were paint cans with paint sloshed over their sides scattered along the walls of the building on top of and under the workbenches. There were greasy rags which must have been used for cleaning. There were stacks of lumber lying all around with pieces cut at so many different lengths that it was impossible for them to be collected into neat piles.

When Mr. Ritchey and I finished our little "get-to-know-each-other" confab, which must have lasted an hour and a half, I was taken out to my first assignment, to begin cleaning and repairing the glass-bottom boats. At this point, at the end of the summer season, four boats were in line for major repair jobs.

While we were walking toward the docks, I had a chance to look over the layout of the maintenance department. The shop where I met Mr. Ritchey was at the end of a canal about one hundred yards long, and the building was ringed by oak trees with vines and clumps of moss hanging down from their branches. The trees with all their moss and the surrounding underbrush hid the shop from the view of visitors to Sparklin' Waters.

There were docks made of creosoted timbers along the side of the canal outside the shop. There was enough space for six glass-bottom boats to be tied to the docks at night for routine cleaning and service. During the spring and summer months, one man began work about dark and spent most of the night washing the windows and seats of the boats and making routine inspections, getting ready for the next day's trips up and down the river. At the section of the docks nearest the maintenance building and connecting them to the main building was a ramp where one boat could be suspended in dry dock. That morning, the dry dock was empty.

After we walked around for a few minutes, Mr. Ritchey gave me my first call to duty. "Newt, I guess we had better get started. We have got to overhaul all the glass-bottom boats. I'll show you how and then help when you need me. We'll clean them, one at a time, starting while they are tied up to the docks but still floating. After the inside of a boat is finished, we'll pull it up in the dry dock and work on the underside."

That first day, we started cleaning the boat nearest to the dry dock. Mr. Ritchey spent all day showing me what to do. After that, he spent an hour or two working with me every morning for the next month.

The first thing we did was take out the floor grates. The deck underneath was covered with sand that formed a shell almost like concrete two inches thick. The grates had to be cleaned with a steel brush. After that, they were scrubbed, washed and dried. Then they had to be painted with a white tung oil-based paint. Sand around where the grates fit in had to be dug, sometimes even chipped, out. Then the decks were scrubbed then sanded with coarse sandpaper, cleaned again, and left to dry. Just cleaning the inside of a boat took all of one, sometimes two, days. Then we had to make repairs. Wooden sections that were broken or rotting were replaced with dark red cypress lumber. Next the inside deck and walls as well as the seats, doors and petitions were sanded and varnished with two coats of a special clear tung oil glaze.

The boats were then moved to the dry dock, where they were cradled in slings and lifted out of the water.

When I towed the first vessel up to the dry dock, Mr. Ritchey helped me to get it slung. Then he showed me how to clean and overhaul the outside of the boat. The crud and debris had to be scrubbed or chipped off the wooden hull and the glass flooring. Old whiskers and moss that had built up below the water line had to be scraped off, and then the hull had to be sanded. After the woodwork was cleaned and sanded, it was painted snow-white with tung oil marine paint. Then the glass was cleaned and polished. Within a few weeks, I got the hang of overhauling the boats and was able to take full charge of repairing the fleet. When one boat was cleaned, painted, and put in good working order, I moved on to another.

Right from the start, it was obvious how excellent a carpenter Mr. Ritchey was and how good he was at painting and plumbing. That came out while he was showing me how to make the repairs. He was not a show-off; he was just a good woodworker and teacher. But how was he as an electrician or a machinist? It was almost a year before I had a chance to see how good he was at putting together new machinery when there was a need.

The glass-bottom boats, plowing the waters of Sparklin' Waters when I started working there, used

Johnson gasoline kickers. Their fuel tanks were refilled at the docks while passengers from an earlier trip were getting off and new passengers were getting on board. During the process of refueling, a bit of excess gasoline often spilled into a trough underneath the tank and ran down into the river, where it was washed downstream. In those days, nobody was concerned about such a small amount of pollution, but one day, while the tanks were being filled, a lot of gas spilled onto the deck. I heard that the pilot had been drinking although I never knew that for certain. When he pulled the rope to crank the engine, it backfired, and the whole boat caught on fire with twenty-six passengers on board! We lost only one. Unfortunately, a man who was standing on the pier at the time of the fire jumped into the river to try to save his two children from the boat who had jumped into the river. He couldn't swim, but he dived in despite warnings from everybody standing around him. That included some of the staff as well as his relatives. When he dived in, he sank to the bottom of the river before we could get to him. His children were rescued.

  It was about a week after that disastrous fire that Mr. Mack Tracy walked into the maintenance shop looking for Mr. Ritchey. When he found him, he asked, "Dick, would it be possible to replace the gasoline engines on the boats with electric motors?"

  "Well…, it might be. Let me think about it for a day or two and I'll see what I can come up with."

  Mr. Ritchey didn't say very much to the rest of us working with him over the next few days. He had a little note pad that he kept scribbling on. He took one of the gasoline kickers and sketched out some plans. He measured the distances between the holes where the engine was connected to the drive shaft, he measured the size of the holes and then made a sort of blueprint. After doing all of that, he drove over to Jacksonville and found a shop where electric motors were being used to power hoists. He bought one of the motors, and before another week was out, he sketched a plan to seat the electric motor over the drive shaft and connect the shaft of the motor to the shaft of the Johnson kicker. In less than two weeks, he had diagrammed a system to replace the gasoline engines. Electric motors were put on every one of the glass-bottom boats, and there was no more fire hazard. The electric motors used the same motor mounts. Four bolts connected the shaft of the motor to the drive shaft that turned the propeller. A rheostat on top of the engine regulated the flow of electrical current and the speed and direction of the rotation. There were three speeds in each direction. The pilot simply pushed a lever back and forth to regulate the speed and the direction the engines and propellers turned.

    Huge storage batteries were used to supply power. The first ones were so heavy that it took two husky men to carry one. Eight fully charged batteries stored enough current to run the motors for ten hours. One battery was placed underneath the deck on each side of the boat midway back, and six batteries were placed in the bow. The batteries were wired in such a fashion that all were charged overnight by a single cable connected to the service docks. The battery posts had to be coated with resin and Vaseline to keep them from corroding. Since that time, the glass-bottom boats have been remodeled, but power to drive them still follows the plan Mr. Ritchey worked out.

AFTER I SPENT the fall and winter overhauling glass-bottom boats, my next big job was to paint signs advertising Sparklin' Waters. For those, we cut wooden arrows, six feet long, six inches high and one inch thick. The signs were painted silver with blue lettering that read, "See Sparklin' Waters in Florida."

I cut out the boards, then coated them with aluminum paint and stacked them, one on another, as high as my head. Later, Mr. Ritchey helped stencil them. For nearly a year, I painted and stenciled, what seemed like a zillion signs for posting along highways from the Great Lakes to Florida and from Texas to the Atlantic seacoast.

Bud Mayson, Shorty's brother from High Springs, and Johnny Small, a hired man, and I stacked the signs on the back of a Chevrolet pickup truck. We added two kegs of sixteen penny nails, hammers and a stepladder they could use to hang the signs. Mr. Mayson and Johnny drove the load of signs to Sault Ste. Marie, Mich. There, they started posting signs and kept doing it until they got back to Sparklin' Waters. Along the way, they nailed signs on tall trees, on fence posts and on buildings as high as they could reach from the ladder. They tried to place the signs in clear view of travelers and get them high enough so that a mischief-maker could not reach them unless he had a step ladder. The arrows always pointed in the direction of Sparklin' Waters.

AS LONG AS I worked with Mr. Ritchey, I did whatever he asked. Besides cleaning and repairing glass-bottom boats, we cleaned out clogged sewer lines, repaired electrical circuits, repaired heating systems, painted buildings or parts of buildings, and replaced hinges and locks. We replaced rotten wood on fences and buildings as well as on the boats. We did whatever was necessary to keep Sparklin' Waters in working order and looking attractive and shipshape.

I was sent to work with Mr. Ritchey at a time when I needed more education. I needed a teacher, but I also needed someone I could trust and respect. I needed reassurance. Mr. Ritchey filled those needs, and for what he did for me, I will always be grateful. He was my teacher. He was the greatest practical teacher I ever had. I always listened to his instructions. I learned a lot from him.

I WAS MR. RITCHEY'S pupil and helper for two and a half years and thought that working as his assistant was a permanent job. It came as a big surprise to both Mr. Ritchey and me when Shorty Mayson walked into the workshop one Friday afternoon and told me, "Report to Mr. Masters near the parking lot beyond the circle in the driveway at nine o'clock next Monday morning."

The valet had quit and I was assigned to take his place. That move ended my student days, but no assignment could erase, or even reduce the admiration and respect I had for Mr. Ritchey. After I was given my new place to work, Mr. Ritchey and I met and talked once or twice a week as long as I was at Sparklin' Waters.

5

"**NEWT, GO OVER** to Reinhart's Clothiers on the square in Okena and get measured for a uniform."

That was my greeting when I walked into the parking lot where Mr. Masters was parking cars while he was waiting for me. He was a tall, wiry gray-haired man in his late sixties. He owned the land around the head of Sparklin' Waters River and leased five hundred acres to the Tracys four years before I started working there, but he was interested in seeing the resort become a nationwide attraction, so he spent his time helping with the management. That Monday, he was parking cars for the customers.

I expected to start parking cars as soon as I got to the lot. I thought I'd spend my day parking cars and then run to get them when the customers were ready to leave. Instead, I spent my morning walking—almost trotting—into Okena. It was about ten o'clock when I got to Mr. Reinhart's store. He had just opened for business.

"Good morning. I'm Newton Sexton. Mr. Masters sent me over here to be fitted for a uniform."

"Oh yes. He told me he was going to send you over."

After that, he took me to the back of the store and pointed to a square piece of carpet in front of a mirror with panels on each side.

"Stand here while I measure you."

With that, he pulled out a tape measure and stretched it from the floor to the top of my head. "Six feet-four. Kinda tall for a kid."

He jotted down the numbers on a piece of paper, then reached around me and measured my waist, then my chest, even my neck, and finally, my head.

"Spread your arms out." Then he stretched the tape from my spine to my wrist and then from my armpits to my wrist and again wrote something down.

Next, was from my waist to the tops of my feet. Afterward, he ordered, "Spread your feet." And he reached up into my crotch and stretched the tape to the tops of my feet again. When he did that, I jumped. I thought he was about to measure my privates and that was where I was going to draw the line, but I guess he got what he needed.

By that time, I thought he had measured everything he could think of, but there was more. He had me sit down, and he measured my feet, heel to toes and the width from the inside of the balls of my feet to the outside of my little toes. I don't think he could find anything else to measure. That all took about half an hour.

When the measurements were done, Mr. Reinhart told me, "Come back next Monday. I should have your suit ready for you by then."

"Thanks a lot."

With that, I started my trek back to Sparklin' Waters. I got to the parking lot just about noon and found Mr. Ed still parking cars. I took over the minute I got there and parked cars in my scrubby clothes all the rest of that week. The next Monday, I made another walk into town to pick up my uniform. When I walked into the store, Mr. Reinhart recognized me right off and handed me two suits and a pair of shoes. He reached in a hat box and handed me a cap.

The shirt and uniform coat and pants were white. There was a little gold bow tie to be worn on the shirt. The coat was double-breasted and had long sleeves with gold braid above the wrists. It also had large brass buttons in front. The pants had a matching stripe of gold braid down the side. The captain's cap was simple. It had a black headband and the top spread out above. The brim was white with a line of gold braid across it. The white shoes had thick soles, but the edges of the soles were painted white so their thickness not noticeable.

Once I was dressed in my monkey suit, Mr. Reinhart stood back and grinned. His pride was visible. He stuffed my scrubby clothes in a paper bag and handed it to me. Holding the bag and the coat hanger with my second uniform, I started my trot back to Sparklin' Waters. By the time I got there, I had to slow down to keep from getting too hot and sweaty and wearing blisters on my feet with the new shoes.

When I got to the parking lot, Mr. Masters was smiling just like a chessy cat. He was obviously just as proud as Mr. Reinhart was to see me in my new monkey suit.

"Newt, I like that uniform. It really makes you look official."

I didn't say a word. We had a difference of opinion. It may have made me look official, but it sure made me feel like a dunce. I just didn't like it right from the start. I felt like I was in a cage.

With that, Mr. Masters got down to business. "I think you need to pick a few pansy blooms from the flower garden every time a couple is getting ready to leave and offer them to the lady. When you hand them to her, give a little going away speech like, 'I hope you had a pleasant visit. As a remembrance of your stay at Sparklin' Waters, I want you to have this little bouquet.'"

"Yes, Sir!" I answered, just like I always did when I was told to do something.

I was impressed with the response when I gave the ladies the bouquets. When I presented those pansies and gave my little speech, I almost always received a tip of at least a dollar, sometimes as much as five bucks.

While working as a valet, I learned another lesson which has stuck with me through the years. One day, Captain Shorty happened to be walking by when there were two luxury cars parked in the driveway, one behind the other. When the first limousine pulled away, the driver in the second car asked, "Who was in the car ahead of me?"

"Um…I'm sorry. I don't know." I guess I was a little embarrassed and I may have shown it.

Shorty overheard the conversation and, after the second car pulled away, called me over and warned me. "Don't ever remember who a person is when you are questioned."

I've never forgotten his advice. I guess the good Lord was looking after me at the time, since I had to say I didn't know the people in the first car

I liked being a valet. I made tips. Lots of times they were big, just for parking cars for well-heeled customers. The only thing I didn't like was having to wear that monkey suit.

I stumbled on another perk soon after I started parking cars. Tom Cheny was manager of the Sparklin' Waters Grille. One day, he asked me, "Could you send some of the guests over to my diner? I'll make it worth your while. You can eat free there so you can tell all the customers what the food's like."

From then on, I paid for very few of my meals, and they were surely more scrumptious than those I was fixing down at Camp Nasty. However, I had a little surprise the first time I went to the grille. When I opened the front door to go inside, Tom met me and said, "You should come in through the back door. That's where all the workers at Sparklin' Waters come in. That way, you'll feel like part of the family."

That made it clear to me. I was to be treated just like all the other hired help. I didn't like it much, but who could complain when the meals were so good and, besides that, free?

It was just a year after I went to work with Mr. Masters that he came out one Friday morning and announced, "We need for you to take over as Captain of the Boats. We need a big man for that job, one who can help passengers get on and off the boats and one who commands the respect of the pilots. I'll get somebody else to be a valet, but we need you to go over and start managing the boats next week."

"That's fine. But what do I do? I've never done that sort of thing before."

"Oh, I know that. All you have to do is to stand at the loading docks and catch the tie-lines when the boats are being docked and be sure that ropes are tied tightly to the pier. Then take up tickets and load passengers for the next trip down the river. Some will need help. You should get the passengers loaded as quickly as possible, but be sure that no more than thirty get aboard each boat. Once all the passengers are on, untie the lines and throw them over to the boat captain. Then wait for the next boat to come in. When that boat gets close to the wharf, the pilot will throw you the bow line. Tie it to the post. Then he'll throw you one from the stern and he'll go back to the helm and turn the rudder sharply and rev up the engine. That will push the stern closer to the dock. Tie the back line. You may have to retie the bow line just to make sure there is no space and no possibility of a space opening up between the side of the boat and the pier. Then you'll push the gangplank onto the boat and unhook the chain on the railing. As the passengers get off, it would be nice for you to give them one of your little greetings, pretty much like the goodbye speeches you've been giving the ladies when they are leaving Sparklin' Waters. That'll make the passengers feel good. Always offer the ladies a hand both when they are getting on board and when they are getting off."

Over the weekend, I made up my new little speech: "I hope you had a nice ride. I hope you saw a lot of interesting things through the glass window. Please come back for another trip down the river at Sparklin' Waters real soon."

As soon as all the passengers got off the boat, I ushered new passengers over the gangplank. When the boats were loaded once again, I hooked the chain across the railing and pulled back the gangplank. Then I untied the lines and threw them back on the boat. A new boatload of explorers was on its way.

As Captain of the Boats, I had a lot more responsibility than when I was working as a valet. I liked that, but there were two things I didn't like about my new job. Mr. Ed still wanted me to wear the monkey suit and there were not many tips. I really liked the tips I had gotten while working as a valet. I started looking for ways to step up my pay.

The pilot of one of the sightseeing boats came up with a scheme for getting the passengers to hand over more tips without suggesting that they were wanted or even expected. A cotton rope was rigged in the ceiling of the main deck around some quarters lying on a piece of plywood. It was out of sight of the passengers. When the cord was pulled, a quarter fell through a hole in the ceiling just like pennies (twenty-five of them) from heaven. The money dropped onto the glass plate at the bottom of the boat, making a clanking sound when it hit the glass. Just as the quarter went "kaplunk," the pilot called out just loud enough for any passenger to hear, even if he or she were not listening, "Thank you, sir." or "Thank you, Ma'am." "Just drop it on the glass. It was so nice having you aboard."

The sound of money falling on the glass stirred up lots of donations. Sometimes, the pilot would have to drop two quarters before money started rolling in, but donations would come almost always. Every once in a while, some kid would grab the money and run off the boat, but that didn't happen very often.

Another method of increasing my income cropped up when visitors got to the ticket office too late for the scheduled tours. I gave them personal cruises down the river through the wonders of Sparklin' Water. Grateful explorers gave me big tips for the special attention they got. Even though they were late and we rode in my open canoe, I could show them all the wonders they might have seen from the glass-bottom boats.

I rode along on a lot of the tours myself. I sometimes caught catfish for Aunt Kate. To catch a fish, I put a chair on the deck next to the boat captain. I sat there with a wet towel spread on the deck between my feet. During stops along the way when the pilot was explaining what could be seen through the glass or just taking time for the passengers to look along the riverbanks, I dropped a three-foot fishing line next to the idle propeller. My hook was baited with bread. When a catfish snapped it, he was caught. I hauled in fish as fast as I could lift the pole.

To keep the fish from flopping around and catching the passengers' attention, I rolled the fish in the wet towel I had spread on the deck. Four to six fish could be rolled into a single towel. Since the passengers were so entranced with what they were seeing underneath the boat, nobody ever even noticed my catching fish. I took the fish to Camp Nasty where I sometimes had fish fries or I took them to Aunt Kate's where she cooked feasts.

"HOW 'BOUT LET'S have a fish fry on your next day off? Let's have it out at your place—at the cabin?"

That request was made by Aunt Kate's cousin Blufton, a chief petty officer in the Navy who was visiting while on leave. He grew up in Dupont, Ga., and was visiting Okena just for the weekend.

"Okay, I'll catch some good fish, and we'll cook 'em at Camp Nasty!"

I caught what we needed. Aunt Kate drove up about ten-thirty in the morning and brought a bunch of the kinfolks with her. They came the back way on a little dirt road. Blufton and I were sitting beside a high worktable I had gotten for just this purpose. We cleaned the fish and put the heads and entrails in a fifty-pound nail keg with plans to take them to Sparklin' Waters that afternoon when I went back to work. I always gave the fish cleanings to the two pet raccoons everybody played with. Those coons walked around the place just like they owned it!

Before I was ready to leave for work, I noticed the coons prancing around my yard. I opened the keg, and man, that macho coon let me have it! Both Blufton and Aunt Kate wrestled with the coon trying to get him to let me go. He bit me, up and down my arms, on my body, on my legs—everywhere. He just ate me up!

"You son of a bitch. I'll kill you if it's the last thing I do!" I yelled as I grabbed my rifle and shot at the coon, but I missed, or if I hit him, he didn't show any signs of it.

The bastard tore through the woods and I hotfooted it after him with my gun. Every time I saw the coon in a clearing or climbing a tree, I took another shot.

Finally, I killed the varmint. Still splattered with blood, I grabbed the carcass and carried it over to Rob Allen. I explained, "This damn coon bit me all over and I want you to save the hide for me."

It took him about two weeks, but Rob made a nice rug of the head with the hide still attached. When I went to get it, he instructed me, "You take this rug and put it in front of your fireplace!"

I spread it out just like he recommended. Wherever I put the pelt, its two eyes seemed to stare right at me!

I kept it for years, but when I got married, I buried the coon. In the meantime, I cussed him every day!

6

"NEWT, WHAT ARE you doing tonight?"

Shorty Mayson walked up behind me and asked that question when I was standing on the docks waiting for another glass-bottom boat to come in to be reloaded with passengers.

"Nothing, as far as I know now."

"How about getting your canoe out and cleaning it up real nice. I have a lady who wants to take a ride down the river. She's a star on the Grand Ole Opry, and I'd like for you to give her a special tour. Here's ten bucks. You need to get some cold drinks, a few sandwiches, maybe some potato chips or peanuts, or whatever. Just show her a good time."

"Yes, sir." My answer was just the same as always whenever Shorty gave me orders. "But what time should I expect her and where?"

"I'll bring her down to the sand beach about five-thirty."

"I'll be there and I'll do my best to show her a good time."

With that, Shorty turned and started walking back toward his office.

I asked Jonathan, a lifeguard who was not working at the time, to take my place at the docks while I ran over to the sand beach and paddled my Old Towne Red canoe up to the dry dock. There I got a hose and washed off all the dirt that had collected both inside and out. I also got a steel brush with a scraper on the back and scraped and scrubbed off all the mud that had collected under the seats. Some had hardened just like cement, and it was hard to get it all out. Then I paddled the boat back to the sand beach and turned the canoe upside down so that it could drain and get completely dry. After that, I walked over to the Sparklin' Waters Grille and asked Tom Cheny to make six sandwiches two chicken salad, two tuna salad and two ham and cheese. Tom put them in an ice chest along with several bottles of Coke, two bottles of root beer and two bottles of Orange Crush. I also bought four bags of potato chips and two bags each of boiled and parched peanuts. I took the groceries back to the canoe which, by that time, was not only sparkling clean but dry. I put the ice chest in the bow after I turned the boat upright.

It was almost six when I looked up and saw Shorty Mayson come marching across the grassy lawn toward the sand beach holding the arm of Laura Evans beside him. She really looked great, wearing a white satin blouse and a tan-colored gabardine skirt. She had a bright red cloth sash that looked like a belt tied around her waist. Her feet were bare except for some tan-colored sandals that matched the skirt. As they got closer, I could see that she had on thin tan stockings, which almost hid her bright red toenails. This day in time, she wouldn't be considered dressed for a ride in a canoe, but it was rare for women in those days to wear slacks. The star's sandy colored hair was pinned back and held in place with a kerchief that matched her belt.

"Gee, what a gal!" I thought to myself, but I dared not

open my mouth lest I be heard saying the wrong thing. So I just stood and waited until they got to the sand beach. Then I smiled and nodded.

"Newt, this is Laura Evans. I know you have heard her sing on the Grand Ole Opry. I've seen you listening to that program on WSM."

Shorty kept on with the introductions. He turned to the star and said, "Laura, this is Newt, who has been with us three or four years. I can vouch for him. He is an excellent canoeist, and also he's an excellent swimmer. Besides being a lifeguard here, he's been swimming in front of cameras for the last year, so he can take care of you if anything should happen. He's also working with the movies. I really don't know what he does, but it must be important. They're always asking for him. Maybe you can make him a star."

After the introductions and a bit of idle chitchat, Shorty turned to Laura and added jokingly, "I have to warn you. Newt is a real Don Juan. He's always escorting girls down the river. He's young but old enough. He's real macho."

Laura laughed as she screwed up her face and stepped over into the canoe. As she was getting aboard, she replied, "I think I can handle him." She looked like she had muscles enough to handle most any man.

By then it was time for me to take charge. I surely didn't want anything to go wrong. "Please sit here in the middle. I have to warn you to keep seated. If you stand up, you could tip the canoe."

"Gee! I sure don't want to do that."

She followed my instructions to a "T." She took her seat in the middle of the canoe facing the bow. I pushed the off the sand beach, took my seat in back and began poling toward the jetty where the glass-bottom boats were tied up. We waved goodbye to Shorty, as we started downstream. I followed the usual route of the glass-bottom boats, pointing out the sights of significance.

Near the docks I pointed toward the floor of the lake, telling Laura, "This is Mammoth Rock, and here the water is seventeen feet deep." I paddled only a few feet down the river and pointed out the ledge of the rock and the change in the appearance of the water and told her, "The depth of the river bottom drops from seventeen to thirty-one feet. Look at the white sand at the bottom."

Poling a few feet farther downstream, I circled Big Spring and Endless Cavern where the Floridan Aquifer emptied through the giant underwater cavity. I explained, "The floor of that cavern down there is fifty feet below the surface, and enough water flows into the Sparklin' Waters River to keep it flowing at a rate of four miles an hour." I was also able to tell her about finding skeletal parts in the sand far below the surface and diving down to extract a tooth and the whole skeleton of at least one prehistoric animal. I told her all I knew about the animal, explaining that Shorty (I called him Mr. Mayson.) had it mounted in a glass cage which was ready to be placed in the ticket office.

I poled on to Reception Hall where sun's rays were still bouncing off the walls. They were almost sparkling. I pointed to the petrified skeleton of a dinosaur measuring an estimated one hundred feet in length. Fish were swimming all around it. The water was really blue. She gulped and called back, "Gee, it almost takes my breath away!"

I poled down the river, floating over some crevices lined with a flimsy covering of shells. "This is the Bridal

Chamber. They tell me it's eighty-five feet down there," I said. With that, I handed Laura a few bits of bread and commented, "Here, you can feed the fish. If you hold the bread in the water, they'll come up and take it right out of your hand."

She took three large hunks of bread and held them down below the surface of the water, one at a time. Scads of catfish came right to the surface and grabbed the bread out of her hand. Laura squealed, "Gee, that tickles."

She let one fish get too close, and it bit one of her fingers, but she drew back as she screamed, "Ouch!" I looked at her hand but saw no evidence that her skin was broken. She kept feeding the fish until all the bread ran out.

It was while the fish were eating that I mentioned the "Legend of Aunt Lila." I told her that I would tell her the story when we stopped to eat.

When we started our tour, I was a little nervous and I know I must have sounded sort of stuffy. Laura broke the ice by praising me. She said, "Newt, you sure know a lot about what's in the river and what's on the bottom, but maybe you should be a little more at ease." She was so easy to talk to that I began to feel better. In fact, she may have even stimulated me a bit. I felt a little twitch in my lower belly but quickly suppressed that. I was really having a heyday chauffeuring a real superstar from the Grand Ole Opry around in my canoe.

We continued our tour down the river to the Ladies' Parlor where a thousand rainbows reflected against a background of alabaster pink. I poled on to Devil's Kitchen so that she could see the bubbling from the murky bottom, and I pointed out that it looked like smoke coming off campfires.

It was in the Geyser Springs that the Icthyosaurian remains were found. Those were the skeletons of some prehistoric reptiles that are now extinct. At the Blue Grotto, the water was almost heavenly and clear as a crystal for as much as forty-one feet down. Fish paraded through the water in military formations.

We kept on to the decorated Christmas Trees and saw huge fern-like growths, twelve feet high, reaching from a depth of forty-five feet. Scattered throughout the branches of the trees, snails were latched on to the leaves of the fresh water grass. Many think an old boat sunken in the area was lost by some of Hernando de Sota's men. The boat was carved from a cypress tree and was then petrified.

I poled the canoe on to Turtle Village, where turtles, some gigantic, or at least I thought they were, could be seen climbing on limestone rocks sticking out of the water. Laura took some snapshots with her box camera. I never heard how they turned out, but there were lots of turtles, all sizes, swimming around. They ranged in size from that of a small saucer to a couple of feet across. They moved so slowly that she surely could have gotten some good pictures.

The Garden of Eden was filled with brilliantly colored coral and ferns. They looked just like waxed bridal wreaths. All that added to the attraction of Sparklin' Waters.

I stopped at the Legendary Indian Cove, where I retold the old story about the Indian lovers, Oklawaha and Winonak, whose love affair was tormented by tribal hatred. According to the story, they dived down into the waters, trying to get to the land of their ancestors. They were never seen afterward.

At Fisherman's Paradise the river was a little wider and was home to thousands of fish and turtles of all different sizes. They came up to the boat like they wanted to play games. I gave Laura a few more hunks of bread so she could lean over the edge of the boat and dip them into the water with her bare hands. Hundreds of fish swam up to her and she screamed out when they tickled her fingers.

Water was always boiling out at the Volcanic Basin. To me, it looked just like a volcano, at least what I thought a volcano looked like. Large springs spewed out charcoal shells and ashes.

At the Catfish Hotel, fish, some weighing as much as fifty or sixty pounds, came up after more bread. It seemed like they would eat it by the loaf. I repeated the old yarn about the hotel being full of guests and every having room running water.

We drifted over the area filled with garfish. I warned her to keep her fingers out of the water, and I pointed out that those fish have long snouts or beaks and rows of long razor-sharp teeth on each side of the jaw. They look a lot like swordfish and are dangerous. Garfish can weigh as much as one hundred pounds but they don't taste very good.

By that time, it was nearly seven o'clock, so I poled the canoe down to the jetty in front of Camp Nasty and told her, "We can sit on this pier and have our supper. I hope you like sandwiches. We have some soft drinks and potato chips to go along with them."

"Oh, that's great."

We climbed up on the dock, and I ran up to my place to get some folding chairs. When we sat down, she commented, "Oh, this is nice. What kind of sandwiches do you have?"

"There's chicken salad, tuna salad and ham and cheese. Which do you prefer?"

"All that sounds good. Can I have half a chicken salad and half a tuna salad?"

I opened the packages of the chicken and tuna sandwiches and let her take what she wanted. I ate the ham and cheese—both of them. She washed down the sandwiches and potato chips with the Cokes, and I drank an Orange Crush.

7

IT WAS WHILE going through some old brochures of Silver Springs, Fla., that I found The Legend of Aunt Lila printed anonymously. I liked it so much that I memorized it but adapted it to the setting of Sparklin' Waters. I thought that Miss Evans might like it as much as I did, so while we were eating, I told her the saga.

ACCORDING TO THE story, Gen. Henri Dumas was a wealthy Southern gentleman who owned an old plantation not far from Sparklin' Waters. His plantation stretched across many fertile hills as far as one could see. He was an old blue blood, who took back his prewar elegance after the South lost the Civil War.

Gen. Dumas kept tight control on all his hired help, and he tried to control his only son, Pierre, whose temperament was quite similar to that of his mother, who had skipped out long before and gone to live with her sister. Pierre tried to separate himself from his father's oppression by being more and more involved in hunting and other outdoor sports. He spent most of his days and many nights in the woods and swamps around Sparklin' Waters.

One day when Pierre was resting on the banks of the river, he heard some leaves rustling and turned just in time to catch sight of what he thought, at first, was a blonde fairy. She disappeared along a path running in the direction of a shack which he later learned was the home of Aunt Lila, an elderly black lady with both healing and bewitching abilities. He tried to run after the blonde fairy but was exhausted by the time he reached the cabin, and the image he envisioned was nowhere to be seen. Aunt Lila came to the door when he knocked, but she denied having seen the beauty he described.

After two other similar sightings, he realized that this woodland nymph was not a mirage but a real woman, one with whom he fell hopelessly in love. He learned that she was a beautiful young lady full of life. Her name was Ruth Barnes. She had recently moved from DeLand to make her home with a cousin in Okena.

Aunt Lila had taken Ruth as her adopted child when she nursed her through a long illness. Afterward, Ruth spent hours and days at Aunt Lila's cabin. Aunt Lila read her fortune in the tea leaves, telling Ruth of a future of wedded bliss in that big colonial home up on the hill with a handsome young prince who looked very much like Pierre Dumas. Aunt Lila was really struck with Pierre.

Pierre was able to talk Aunt Lila into arranging for him to meet Ruth. Their meetings soon developed into intense love and devotion, which they shared in Aunt Lila's shanty or on the banks of Sparklin' Waters, where they would sit and watch the rippling waters for hours and dream about a future life together.

When Ruth promised to become his wife, Pierre gave her a little bracelet that he brought to surprise her and to seal their engagement. At the time he didn't have a ring. Theirs was an intense love in which they were very happy, and they promised that it would last for the rest of their lives and that no force could ever separate them.

Unfortunately, Pierre's father was intent on his son marrying above the means of poor Ruth, and he managed to send Pierre to a far-off country when he heard of the engagement. He hoped to have Pierre forget Ruth and marry someone nearer his station in life.

Before Pierre left, he promised to write every day and come back to claim Ruth as his wife. However, Gen. Dumas managed to prolong Pierre's stay. He even intercepted the letters of both Pierre and Ruth. Days stretched into months, and eventually almost a year went by without Ruth getting even a note from Pierre. She was heartsick and grieved so much that she stopped eating.

Even Aunt Lila was not able to deal with the skinny figure that knocked on her door. Realizing that she was going to die, Ruth crawled into Aunt Lila's cabin where she and Pierre had spent so many happy hours. There, lying on her deathbed, Ruth made Aunt Lila make a promise, so weird and awful that the old black lady quivered and pulled her shawl close around Ruth's neck. Then she planted a kiss on the feverish head of the dying girl whom she had once nursed to health, but for whom she felt helpless to heal or even help now. In the dead of night, pitch black except for two twinkling stars and silent except for an occasional hoot of an owl, Aunt Lila hauled Ruth's body, bound in a sack, to her skiff, which she had tied to the broken-down pier in front of her cabin. With all the love and tenderness she could muster, Aunt Lila slid the body into her boat and rowed slowly to the Boiling Springs, where she lowered Ruth's mortal remains into the rocky crevice below. That satisfied her commitment, horrible as it was.

Pierre returned the next day, the date set more than a year earlier when he and Ruth were to be married.

He mourned that he had not heard from Ruth during the whole year even though he had faithfully written as he had promised. He wondered whether she had found another lover while he was gone, someone she had revered even more. His father had told him that women were fickle that way. He wanted to take just one more look into the Boiling Springs that he and Ruth had loved so much before he went home to court the rich cousin his father had selected to be his wife.

He found Aunt Lila sitting in her boat tied to the pier in front of her shack, where she had remained throughout the night. Her eyes were downcast, and she hardly raised them when Pierre greeted her as he slipped into the boat and started paddling toward the Boiling Spring. When he was able to anchor the boat over the crevices, he looked down between the rocks eighty feet below.

With horror he saw a woman's hand sticking out above the rocks, and through the crystal clear water, he saw on her wrist the bracelet he had given Ruth.

Without a thought, he dove into the deep water, even though his ears and lungs began to feel like they were bursting. He forced himself to the bottom and into the crevice to grab Ruth's arm. He tried in vain to raise the dead body, but it was caught. Try as he would, he could not free her corpse.

Pierre swam farther to lie beside his dead betrothed and clasped her body to his in a deep and everlasting embrace. Aunt Lila swore that when he did so, the rocks opened to receive the bodies of these unhappy lovers into the bosom of Mother Nature and closed over them. Their bones are still there.

AUNT LILA WAS a fixture at Sparklin' Waters. Her

presence went back as far as anyone could remember. She lived in her cottage and hobbled about the spa all day long. She repeated the tragic story of Pierre Dumas and Ruth Barnes to everyone who would listen: and although she must have repeated it thousands of times, her account never wavered. Before her death, Aunt Lila claimed to be in her one hundred and tenth year. She died a few years before I took her old home for myself. Because the house was so dilapidated and so unkempt at the time, my boss, Captain Shorty Mayson, named it "Camp Nasty."

When I finished retelling the story to Laura Evans, I pointed out that the shanty next to the jetty was where Aunt Lila had lived. I told her that this was my home now. Of course, she had already guessed that, but I just had to tell her how it had gotten its name.

Laura was exhilarated by the tour, by the natural caverns, the underwater growth and the fish, but she was most impressed by my recitation of the Legend of Pierre Dumas and Ruth Barnes. She had even begun to shed some tears as I got to the end of the legend.

When we were canoeing back to Sparklin' Waters, it was still light enough for me to point out the ribbon grass on the floor of the river. This grass was growing in long fibers that waved underwater as the canoe glided past and bent the taller blades. Tiny white flowers made the grass appear like miniature jasmines.

I walked with Laura to her car in the parking lot after we got back to the sand beach. As I opened the door for her to get in, she planted a big, juicy kiss on my cheek and then another on my lips. I never knew exactly what she told Shorty, but I figured it must have been pretty good since only two weeks later Shorty asked me to show Gloria Vanderbilt around.

Miss Vanderbilt came to bring the Rochard Sisters to Sparklin' Waters. They were there to make underwater movies. Since I was responsible for positioning the tank for making movies, I chauffeured them wherever they went. They stayed for over a week, during which time I drove them in Miss Vanderbilt's Pontiac convertible. I took Miss Vanderbilt on another tour of Sparklin' Waters, just like the one I had made with Laura Evans. I just didn't get the kisses planted on my face that came after I took Miss Evans on the tour.

My reputation as a lady's man must have caught Shorty Mayson's attention. He had a very pretty teenaged daughter who caused me to drool whenever I saw her walking around the park in shorts and a slipover shirt during the summer. Although I only spoke to her a couple of times when we passed, Shorty made it his business to warn me not to make any advances toward Mary Jane.

8

"**NEWT, I WANT** you to go down to the Florida State Fair in Tampa on Friday. I'll give you a pass and get Jake to drive you over there in the company car," said Shorty.

"Gee, that sounds great. What do I have to do?" I asked.

"We have a big pile of fliers stacked in my office. You've seen some like them over by the ticket office. You gotta take at least two reams—no you better take at least four reams of fliers that tell about the sights at Sparklin' Waters.

"When you get to the entrance gate of the fair, show the ticket seller your pass and she'll give you a badge that'll get you through the gate. Go in and look around until you find Waxo the Magician and Ida Mae Cook. They have a booth over near the shooting galleries shaped like a caboose. They take turns putting on shows at the back. It's sponsored by Sparklin' Waters and the Florida Motor Lines. You'll represent Sparklin' Waters and there'll be somebody to represent the Florida Motor Lines. I believe his name is James Janson.

"You need to be out front (that is, behind the caboose) and stir up the crowds. Stand right next to the stage and tell about the wonders at Sparklin' Waters between acts. After you get the people to listen, pass out the fliers. Your job will come just before and between the shows."

Those were my only instructions the first winter I was at Sparklin' Waters. The fair was held the last few days of January and the first week of February every year.

When I got Shorty's orders, I started looking forward to a bash. I could remember Mother and Daddy taking us kids to fairs when I was growing up in Ashburn. The fairs were in Tifton, about 20 miles away. We went on Friday evenings after Daddy got off work. I remember watching young kids ride the merry-go-rounds. We older ones rode ponies. The teenagers rode the Ferris wheel and bumping cars. I remembered seeing some side shows, and I had always wondered what was going on inside, but Daddy wouldn't let us go in so I never found out. I also remembered booths where people threw darts at bulls' eyes. I tried it a couple of times. At first I missed, but later I won and got some little soft dolls that not even my sisters could snuggle up to. I could have bought a lot better ones at the nickel and dime stores.

I remembered Daddy taking us to some farm shows where they had some gigantic hogs and others where they had some beautiful cows that they paraded across the lot and auctioned off. Daddy also took us to one show where cowboys were riding broncs. They roped steers and auctioned off some horses, but I never had enough money to even think about buying any of them, and Daddy was not interested.

In other farm shows, they had bushel baskets of corn, lots of green and black-eyed peas, Irish and sweet potatoes and carrots. Some had bales of cotton but cotton wasn't as big a deal in our part of the state as it was nearer the Savannah River. There were always lots of peanuts. All kinds of produce was grown in south Georgia. We also got to see auctions for pies and bread.

Mother cooked some and got "honorable mention" a couple of times.

Mother went in to see the quilts that women made. My sisters went with her to those shows, but I never went. Even remembering those shows, I couldn't imagine what it would be like going to a really big fair by myself.

The first year I went to the Florida State Fair, I carried a tent and a sleeping bag. I put up my tent right between Waxo's and Ida Mae's. Our tents were at the edge of the fairgrounds in the area where tents for the other workers and performers were staked out. While it was not home, my sleeping bag inside the tent was nearly as comfortable as my cot at Camp Nasty. We had an outhouse that we shared with all the other workers at the fair. Boy, did it ever stink! During the week, we ate mostly hot dogs and I drank mostly Orange Crush. Ida Mae and Waxo drank coffee, but I never did. I didn't like it.

I helped Waxo and Ida Mae put up the stage for opening night. I didn't have a lot to do since they did all their tricks in front of the stage. Waxo's magic was the first act. Before the show started, I got up on the front of the stage and called out, "Come one. Come all! Come see the magician from Sparklin' Waters! Come see him pull rabbits out of the hat. Come see Waxo the Magician make cards disappear. Come up close! You'll be glad you did."

People from all around the area of the caboose squeezed in. They came to see the show in droves. When Waxo came out, within seconds he had everybody within hearing distance in a trance. He attracted all the people who could squeeze into the tiny alleyway in front of the booth. Waxo didn't need anyone to hawk for him. He did one trick after another for about thirty minutes

At the end of Waxo's show, there was a ten-minute

break. I took that time to tell about the good things people could see at Sparklin' Waters. I told about the glass-bottom boats, the fish and the turtles in the river, the caverns in the floor of the river, the petrified boat left over from the Spanish explorations, the swimming area, the flower gardens, the antique shop and the restaurant. I told about the crystal clear water bubbling out of the caverns to form the head of Sparklin' Waters River. I emphasized that Sparklin' Waters was just outside Okena and could be reached by railroad or bus as well as by car.

After I talked for five or six minutes, I handed out pamphlets with pictures of most of the attractions at Sparklin' Waters. Everyone took one. I never knew how many people heard what I said or how much they were attracted to the wonderland I described, but plenty of people crowded around, and surely some must have been tempted by the spiel I put out about the fabulous sights.

Then it was time for me to hawk for Ida Mae. "Come one! Come all! Hear the woman make bird calls!"

With that Ida Mae would whistle some calls or coo or crow, whatever came to her mind. Then she would take the stage by herself and begin to make bird calls. Again, a crowd gathered. I really didn't think the masses were as big, nor did the people who were there seem to be fascinated by her as they were when they were watching Waxo. Still, at the end of her show, I took to the stage again to tell about the sights at Sparklin' Waters.

Ida Mae's show lasted about twenty minutes. The shows were repeated every hour. We started about four o'clock in the afternoon and kept going until just before ten o'clock. We had to get everybody out of the place by ten.

I always counted the hours until the shows closed.

When ours closed, I sneaked off to the All-American show. I got there just in time to see the chorus girls sing and dance their last number. After the show, Cindy and Nell and I trotted into town to get something to eat. Unfortunately, the girls' house mother was really strict. They could only go out if there were two girls and the house mother had to know their escort. They had to be in by midnight. She was really picky, but with my hormones raging, I could understand her fear. Still, we all had fun.

The fair went on for ten days. I worked every day and then I took the girls out every night. It was a lot of work and still a lot of fun.

During the second year, the fire department complained that the crowds were too big. The fire chief threatened to close Waxo's show unless the spectators kept moving. I had to push them along. That was a problem, but somehow I managed to stay out of trouble.

Waxo and Ida Mae argued over who attracted the biggest audiences. They wanted me to take sides, but I wouldn't dare. Somehow I managed to sidestep their clashes, but they kept on. Finally, after three years, they decided to take separate nights. Then we could estimate the size of the crowds. As it turned out, even that was hard to gauge since both of them attracted more fans than the fire department allowed, and the threats to close the show kept coming. I was still responsible for keeping the crowds moving.

The Florida State Fair was held every year that I was at Sparklin' Waters and the shows were still going strong when I left.

9

DOG PADDLING WAS about as much swimming as I was able to do when I went to work at Sparklin' Waters. I was a 15-year-old thin, tow-headed country boy at the time. My cousin Jeb Perry was a lifeguard at the vacation spot during the summer. He was expected to keep the swimming area clear of not only rubbish, but also grass. Litter was a problem, but the thing that bothered the swimmers most was grass growing out of the bottom of the river. It seemed like a new crop grew every day. Grass growing in other parts of the river often added to the appeal, but when it grew under the swimming area, there were lots of complaints. The bathers brushed up against it. Cousin Jeb spent hours trying to get rid of the grass. He would dive down and yank out every sprout he could find. I watched him do it for a few days and then decided that I was going to try to help out.

It was about the end of the first week after I moved into Camp Nasty that I started wading into the edge of the river to pull out sprigs of grass. When I started, I waded out until the water was waist high. Then I squatted down and pulled out all the weeds I could reach. I stayed down as long as I could to grab all the weeds I could get in my hands before jumping up to catch my breath. Later I started heading out into the deeper areas until the water was over my head. After I got a handful of grass from the bottom, I'd jump back to a place where my head was above water. I turned loose the shoots of grass and let them float downstream. I used that ducking method for about ten or twelve days before I started paddling underwater. Then I started swimming. With some lessons from Jeb and practice every day, my swimming got better and better. Besides just stripping away the grasses growing at the bottom of the swimming area, I started swimming for longer and longer distances. My stroke got good enough for me to make the title of "Master Swimmer" when I took the Red Cross lifesaving test the next spring. I also qualified as a lifeguard. By the end of my second summer at Sparklin' Waters, I was able to swim twice a day from Camp Nasty to Sparklin' Waters and back. After I qualified as a lifeguard, I began working on weekends and sometimes in afternoons after I was through with my other duties. By the next year, I could swim four miles down the river and back. I could stay underwater for as much as two minutes and forty-five seconds.

    Lifeguards at Sparklin' Waters sat on top of a tower anchored at the deep edge of the swimming area. That was fifteen to twenty yards off shore. The tower was a twelve-inch square post set in a fifty-five-gallon barrel of concrete which was buried in the floor of the river. A small triangular platform with a seat was nailed on top of the post ten feet above the water. Rails closed in three sides of the platform leaving a wide space for the lifeguard to dive off if necessary. From the tower, lifeguards could keep watch over the whole swimming area. We could rescue swimmers from any section of the river marked off for swimming if they got in trouble.

During the time I worked at Sparklin' Waters, I rescued twenty-three swimmers.

While I worked as lifeguard at Sparklin' Waters, I wore an old-fashioned bathing suit made to cover the whole body from the shoulders to the upper thighs. I pulled the straps off my shoulders and rolled the suit down below my belly button, making it look like a modern-day bikini. When I was on shore, the only other things I wore besides the bathing suit were brogans to protect my feet from stickers and stones. I never wore a hat.

"**MR. NEWT, MR. NEWT!!** Johnny's down on the bottom of that deep spring down yonder, and I can't get him out!"

"How long has he been there?"

"Bout two hours."

"What are you coming to get me now for? I can't get him out alive now."

"We couldn't get him out. We tried!" The kid screamed hysterically.

That call came in one Saturday afternoon about two o'clock when I was sitting on the guard tower. Dick, an 11-year-old boy, came running up to me, calling for help. When I heard the story, I knew that even if he had come to get me as soon as Johnny slipped out of sight, I could not have saved him. To get to me, Dick had to run or swim nearly two miles, and by that time, there was no way that Johnny could still be alive. Actually, it turned out that Johnny's body had settled in an isolated spring nearly twenty-five yards from the river's edge. That was a couple of miles from the guard post. I could only get there using my canoe.

After poling through the pass to the spring, I dived down thirty-five feet to grab the dead swimmer by the nape of the neck. I could bring his body to the surface only after going through all of that.

He was as cold as a cucumber, dead as a door knob.

"GOD AW'MIGHTY, I got trouble now!" It seemed as if I were screaming to myself while I was sitting in the lifeguard tower of the swimming area one Saturday afternoon. I had just heard a flatbed truck loaded with women come rumbling into the parking lot. The truck bore the markings of the Florida State Reform School.

I was in trouble! Whenever the women from the reform school rolled in, I always had to call two or three extra lifeguards.

The young women were squeezed together on the back of a cattle truck. They were held in by ropes tied to four corner posts with stakes in between. The guards at the reform school always loaded as many female convicts as they could on the back of the truck. There must have been between thirty and thirty-five women. When the truck screeched to a stop, the girls ran across the lawn to the sand beach and bathhouse for a swim and a cookout. At least twenty or twenty-five dashed into the bathhouse to put on bathing suits. The big problem was that women from the reform school were man-hungry!

Every time they came, at least one would swim to the far edge of the swimming area and then throw up an arm and yell, "Help!" That was just to make the lifeguard dive in after them. Over the years, I pulled out a bunch of them, but mostly they just clung to me. They squirmed around rubbing their bellies and crotches up against me. It felt good. I always got a little tingle out of it, but sometimes the girls hung on and pinned my arms down so tightly that it was almost impossible to swim. That was dangerous. Those women from the reform school were starved for love.

I had to hand it to the matron. When she caught them grabbing hold of the lifeguard or misbehaving in any other way for that matter, she called out, "Get out of the water and come here. Go sit in that truck where you belong!"

She didn't hesitate. She would restrict them to the truck in a split second. She would do the same thing when they tried to flirt on the beach or grab the food prepared for the cookout. When she ordered her girls to stop misbehaving, she took complete control, and they didn't dare disobey her.

"HI, NEWT. LET'S go joy riding. When can we go?"

They called me "Ever ready" because I was always ready!

I had a special little cove where I took women from town for a swim. It was about one hundred and fifty feet off the river. The inlet, just wide enough for me to pole my canoe through, was surrounded by a dense growth of underbrush in between the cypress trees, so the cove was hidden from view of people riding up and down the river. The cove was fed by a constantly bubbling spring. At the back was a nice sand beach about twenty feet wide. Girls who knew about the cove begged to go riding in my canoe.

"Just a minute." I got my canoe and took one, at most, two of them for a private swim. Sometimes we lolled around on the sand. Sometimes we stripped and studied anatomy. I made lots of exciting discoveries in that little alcove.

SPRINGTIME WAS ALWAYS a blast. There is always a lot of new growth on the trees and underbrush, even in the semi-tropics. Springtime was mating season for many of the wild beasts, and springtime was the time when the girls started sporting their new bathing suits. I liked that.

It was a Friday afternoon when Jacco Keaton came rambling up the driveway in his convertible Oldsmobile, which must have been ten years old. He had painted it red, just like a fire truck, and it had a canvas top. Marilyn Wilson was sitting in the front seat next to him. She was a 16-year-old slightly plump brunette whose face was all painted up with rouge, and she wore a plentiful supply of lipstick. She was all smiles. When I first saw her, I thought she was dressed for some sort of formal event, but when the car got closer, it was obvious that she was wearing a beach dress. Judy Kanto and Alice Ward were sitting in the back seat. They also were wearing cover-ups, for what reason, I couldn't figure out until they climbed out of the car and walked down to the beach.

Jacco spoke for the crowd when he asked, "Newt, is it possible for you to take us all down to your little cove off the river so the girls can show us their new bathing suits? I want to see how they look."

"Sure. Let me see if I can find somebody to take my place in the lifeguard's station and we'll see what they look like."

I called Carl Brandon, a skipper on one of the glass-bottom boats sitting idle at the time. He jumped at the chance to take my place.

I walked across the sand beach, untied my canoe and then motioned for Jacco and the girls to come on down. It was a full load, but they all piled in and I poled the boat right up to the sand beach and we unloaded.

As soon as we were on the beach, the girls started molting. They pulled off their cover-ups. What a sight! All three had on one-piece bathing suits, the most exotic I had ever seen. The suits Marilyn and Alice were wearing were colored with flowers which seemed to be painted in just the right places, a big splotch over each of the breasts and a bouquet over the lower stomach area. Marilyn's suit was strapless—at least it appeared so at the time although I later discovered that it had straps to go around the neck. Alice's suit had straps over each shoulder. The two women really looked just like they were posing for the centerfold of a swimsuit magazine.

Judy's suit was striking, all white, and like Marilyn's suit, it had straps tied around the neck. It was a snug fit. Beautiful!

Each girl was itching to show off her suit, so the three of them paraded up and down the beach for several minutes before I asked, "Why don't you take a dip?"

"Well, we haven't tried that yet, but I guess now's the time."

"That's right. Now's the time."

All three dived in and swam out into the middle of the cove and back. Marilyn and Alice came out of the water looking fine, but we all gasped when Judy walked out of the water. Her all-white bathing suit was almost transparent and showed everything! She looked almost naked. She couldn't get her cover-up pulled on fast enough.

I pulled out some towels from under the seats in the canoe, and each of us lay there in the sand for about thirty minutes, long enough for the girls' swim suits to dry. We then all piled in the canoe, and I poled back to the sand beach at Sparklin' Waters. We had a good laugh. Judy was so embarrassed she took her bathing suit back to the store the next morning to exchange it for a multicolored suit.

"**NEWT, I WANT** to show you my new bathing suit. I heard that Marilyn, Alice, and Judy showed you theirs yesterday. I heard that Marilyn's and Alice's were real smart but that Judy's was a knockout, so much so that she decided to take it back this morning. Too bad. I would like to have seen it.

"Let me show you mine. I'll do it right here," said Justine.

It was about nine o'clock on Saturday morning, the day after the parade at our little isolated cove, and Justine Newman had come running out to the beach. At the time, there were only five or six swimmers out there. Justine had a beach towel wrapped around her. I was sitting on a bench in front of the swimming area.

All of a sudden, she stripped off the towel, threw it at me and asked, "How do you like my new suit?"

"It's beautiful." It was. Hers was a one-piece suit, solid blue with straps tied around her neck. After she got my approval, she dived into the river and swam around the edge of the whole swimming area and back to the beach. As she climbed out and stood up, the straps around her neck broke and the whole bathing suit dropped to her ankles.

Seeing her standing on the edge of the sand beach with her bathing suit covering only her feet, I yelled, "Justine, look at your bathing suit. Pull it up!"

She grabbed it and pulled it back on and held it in

place until she could get her towel. Then she ran off. I think she ran all the way home.

After that, Justine was so embarrassed, she never spoke to me again.

"WOULD YOU LET us make a picture of you skiing through a wall of fire?"

"Sure. What's it worth to you?" I asked.

"We'll pay you five hundred bucks."

"I don't mind. It won't bother me."

Gee! What an offer! A big cheese from Randolph Pictures made the offer after I got the reputation of being a good swimmer.

For that much money, how could I turn it down? No way when my salary was five bucks a week. Until that time, I had not done many stunts, but I felt sure that I could ski through the flames and drop down in the water right after I went through the fire.

When they got my OK, a wire cable was stretched across the Sparklin' Waters River about three miles downstream from where the glass-bottom boats took off. Spanish moss was hung over the cable and doused with gasoline. The wire hung low near the middle of the river so that it was only about a foot over my head when I skied under it on a thirty-six by sixteen-inch surfboard. When the boat pulling me passed under the wire, the moss was set ablaze, so I skied right through a solid wall of flames. Right after I passed through the fire, I dropped the towline and plunged into the river. When I crawled up the bank of the river, the photographers and everybody else watching came running up to ask, "Did you get burned?"

"Naw." I acted just like I did it every day and said it

was fun.

Going through the wall of fire, I had covered my eyes with my left hand, but the fire had singed the hair on the top of my head, my eyelashes and eyebrows and the hair on my chest, back, and legs—all over my body. What a singed pup! The crew and their helpers could not stop laughing. But I made five hundred dollars—not bad for a laborer earning a dollar a day.

10

"CAN YOU MOVE the tank a little farther down stream? A little to the left—now a little closer to that bank."

Jack, the cameraman, sounded easygoing, but my job was to do whatever he asked. As a matter of fact, I usually liked getting everything in position for the cameramen, as much as possible ahead of time.

I grabbed a wrench and loosened the nuts holding the bolts that connected the tank to the pontoon frame and let it rise. When the tank was floating, I paddled the whole contraption downstream to where I thought Jack wanted it. I was dumping in a few buckets of water to weight the tank down when he screamed, "No, not there! A little closer to the bank and a little farther down stream!"

I bailed out the water I had just put in the tank and moved the whole contraption again. I was about to pour more water into the tank and bolt it down when Jack hollered again, "Can you turn the window a little more toward the side of the river?"

After turning the tank, I asked, while gritting my teeth, "Is this where you want it?"

"Yeah. That's great."

Two pontoons were connected by a frame that held the watertight tank with a window on its side. The tank was big enough to hold the photographer and his movie camera. It kept them dry while they were shooting underwater scenes. Water was added to the weight of the metal vat to serve as ballast, so water was removed to float the tank before the position could be changed. Then the tank had to be filled, at least partially, in order to get it in position to be bolted down again. Then all the water had to be dipped out so that the cameraman and his equipment could stay dry while he was shooting pictures. In addition to getting the tank in the right spot, I had to be certain that the window was facing where the action would be—the underwater stage.

I poured in enough water to make the tank sink. Then I bolted it down. After it was locked in place, I emptied it and crawled out of the way so Jack could climb inside with all his camera equipment and film and start shooting. By the time he was set up, I had the top of the tank covered with a black cloth. Then Marty Warren, the director, hailed to the mermaids and mermen to go through their routine. At last, the show was about to begin!

The lens of the underwater camera rested against the glass window on the side of the tank. The glass was half an inch thick and twenty-eight inches square. When the tank was in place for making movies, the top of the window was about two feet below the surface of the water. The photographer stood behind his camera, which was loaded with one thousand feet of film. Using this setup, he could shoot all the gyrations of the swimmers. Pictures were made at one place for a while, and then the tank had to be moved to a new spot.

In order to keep control of all that was going on down at the river, the Sparklin' Waters Resort furnished the tank and scheduled its use. My job was to try to satisfy the film companies as well as my bosses.

For months after I went to work at Sparklin' Waters, I moved the pontoons with their cradled tank from one place to another in the river. I just parked the apparatus wherever it was wanted.

Most of the underwater scenes filmed at Sparklin' Waters were shot by MGM, Fox Theaters and Columbia Pictures, but Kennedy Productions located in Palisades, Fla., also made pictures there. Filming was also performed by sports commentators.

"Hello, Jeb! This is Marty Warren, movie director for Grantland Rice. We want to come down and take some funny pictures in the river. The first clip we want to make will be a robbery in the depths of the river. Depending on how that works, we may try a boxing match and then maybe an orchestra playing underwater. We'll need your people for the shots. Two men for the first clip. Can you get with Newt and meet down at the Okena Hotel at nine o'clock Monday morning?"

That was a phone call Jeb had from Marty, who was in New York at the time. Marty and Jack Eaton, his cameraman, made specialty clips for Mr. Rice. My cousin Jeb was their contact man, and he made almost all of the local arrangements.

Grantland Rice himself didn't come to Okena very often. He just sent his crew. They worked out the details including the kinds and numbers of scenes to be filmed. They did most of the planning in New York, focusing on the sequences so that his men could outline what they wanted and project the number of swimmers they needed. Usually, Marty called Jeb to let him know when they would arrive and what scenes they wanted. Those calls reduced the likelihood of conflicts with other companies making movies. The calls also let Jeb know when we would be needed and what they wanted us to do. He then called each member of the swimming team. Despite Jeb's requests, sometimes the photographers just showed up at the Okena Hotel without calling.

When I first arrived at Sparklin' Waters, two of my cousins, Jeb and Martha, did most of the underwater swimming acts. At the time, I had to learn to swim well enough to be used in the skits. Finally one day, while getting ready for some underwater shots, I put the tank in place for the cameraman and dived in. Jack got a shot of me swimming past the set. The next time an extra swimmer was needed, they let me have a part in the scene. It was not long before the movie men asked me and another cousin, Eileen, to work as a team. Eileen was beautiful, and we worked well together. So from then on, Jeb and Martha made up the first team, Eileen and I were the second team, and another couple, Woody and Elsie, made up a third team. We were all good swimmers and each couple was gaining experience swimming as a part of his or her team. We were able to follow the directions of Marty and his cameramen, so it was pretty much a toss-up as to who would be called. Since Eileen and I swam together often and I was nearly always around when extra swimmers were needed, we were used a lot.

Besides pictures of the teams swimming together, some scenes were made to look as if we were relaxing with fish swimming around us. We were able to attract the fish with crumbs we threw out while we were showing off. Other clips were made of the swimmers diving into the river, and swimmers sitting on logs or among the tall blades of ribbon grass underwater. An old wives' tale had it that if a man put a bloom from underwater ribbon grasses in the heel of his shoe, he would be married within six months. I never tried that. For sure, I had all I could do just to take care of myself.

AFTER THOSE CLIPS, the team from New York came up with a plan to make some slapstick swimming scenes. When Marty Warren and Jack checked in at the Okena Hotel, we met them there. Jeb made the introductions.

"Come on in! This is Marty Warren. He's gonna' show us what he wants us to do tomorrow morning."

I followed Jeb to the door and we went in to get our instructions.

"Jeb, you and Newt are the only swimmers we'll need today."

Then Marty launched into his instructions. "To start, I want to make a 'shoot 'em up,' a real 'old West movie' clip, all underwater. We'll use just two men for this scene. I want each of you to be dressed like a city slicker just like in olden times. You'll wear bow ties on a white dress shirt with the sleeves rolled up, not quite to the elbows. You'll each have on dark pants and little black derby hats. Newt, you'll be the robber pointing the gun at Jeb.

"All of this will look just like a holdup scene in the old movies—Jeb, you'll hold your hands up high.—We need to make a little run-through this afternoon."

AFTER MARTY GAVE those directions, he tossed me a pistol as he said, "Newt, I want you to stand in the center of the room. Jeb, I want you to walk in from the side of the room. On my cue, Newt, you should walk toward Jeb and point the pistol with your right hand while holding your left hand out to grab Jeb's shirt and then his back pants pocket, like you are trying to get his wallet. You'll have to make your moves quickly so we can shoot the whole scene with one breath.

We practiced the scene three times before we all felt like we knew what moves we needed to make and when.

"That's good for now. The real question is how will it work underwater. We'll have one or two run-throughs in the river in the morning before we start shooting. How about it, Newt? Do you think you can have the pontoons and the canoe at the loading dock for us at nine o'clock?"

"They'll be there."

The next morning, I swam up the river, naked as always. After I found my bathing suit under the lifesaver's tower and pulled it on, I swam over to the sand beach and shoved the canoe out into the water. Then I paddled it over to the docks by the maintenance barn where the pontoon raft with its cradled tank was moored. I unleashed the raft after I tied the canoe to it so I could poll the two crafts out on the river. From there, we would float downstream. When I got to the spot where I thought the underwater drama should be created, I turned the pontoons so that the window on the tank faced downstream, and I anchored the pontoon craft several feet from where I imagined the stage would be. I estimated the depth of the river to be about eight feet there. After all of that, I poured enough water into the tank to sink it and tightened the bolts. Then I emptied the tank. I untied the canoe and paddled it back to the docks to wait for the Marty, Jack and Jeb.

I timed their arrival just about right. It was nearly nine o'clock when I looked up and saw them coming, carrying the clothes and the pistol. After they came aboard, I polled back to the pontoon raft where Marty and Jack had climbed on board, Jack clutching the camera and film, trying to be sure that they didn't get wet. The stage was set.

Marty stood on the pontoon frame and watched while Jeb and I dressed and dived in. Jeb swam toward the shore and I, armed with my trusty pistol, swam out near the middle of the river to begin swimming back toward the underwater stage on cue. When I got close to the pontoons, I was able to plant my feet on the bottom of the river and stand. I started walking toward Jeb, whose image was clear even at a distance. I pointed the pistol and reached for his shirt and then the back pocket of his pants just at the time I realized that we were about out of breath. We had been down for over two minutes, and I was feeling the need for some air. Both Jeb and I waved our arms and jumped to the surface.

There were problems when we were working underwater. Since we were not able to talk, we had to swim to the surface to get instructions or to tell Marty whatever problems we had or should look for. Another problem we were having was that we were being swept off our feet by the current in the river. We also had to find out just how much we could do between the times we had to come up for air. We just had to work it out.

When we came up the first time, Marty told me that I started walking from too far out toward the middle of the river. He also wanted me to try to walk more and swim less. That was always hard when the current was pushing me off my feet. Jeb was told that he needed to stand closer to the pontoons. Both of us realized that we had to stand with our legs apart and in line with the flow in the river so the current would not sweep us downstream.

Our next performance seemed worse. We were swept off our feet and had to go back and run through the scene again. This time we were better, but there were still mistakes so we tried it a third time, this time in front of the camera with it rolling. When we came up, Marty called out, "Great! I think we got some good film. Let's do it one more time, just to make certain we got what we want. Then we'll call it quits."

The next session went just like the last one, so Marty pronounced it done. Jack got over five hundred feet of good film in each clip. He even made several still pictures. I still have one of those.

Since our picture-taking session for the morning was over, I began poling Marty, Jack and Jeb back to the passenger dock in the canoe. Along the way, Marty let us know his next plan. "I'd like to see if we can take pictures of a boxing match tomorrow. Could we meet at the hotel this afternoon about three o'clock for a dry run? We need three men for that skit. We planned on having you, Jeb and Woody to be the boxers, and Newt, you'll be the referee.

"Jeb, can you call Woody and get him to come down this afternoon? In the meantime, we'll be setting up the scenery," Mary said.

"Sure. I'll give him a call. I'll do that as soon as we get to the docks," said Jeb.

"Newt, will you have time to get all this equipment put away by then?" asked Marty.

"Yeah. I'll be through before then. All I have to do is to take the tank back to the maintenance docks and be sure we have it ready to go tomorrow morning. I may even have time to get a sandwich."

After I got the crew back to the loading docks, I poled out to the pontoon raft and raised the tank from its resting place. I then poled the raft back to the maintenance dock for the night. After hitching the apparatus to the dock, I scrubbed the window of the tank inside and out and left the contraption clean and ready to for taking pictures.

When I got everything done at the docks, I trotted over to the grill to grab a bite. I had a po-boy and washed it down with a big glass of milk. I had barely enough time to swallow it before starting over to the Okena Hotel.

When I opened the door to the conference room, I gasped. The boxing rink almost filled the room. It was twenty feet square. The steel corner posts were four inches square. They were attached to the metal flooring with steel plates. The floor of the boxing ring was covered by a canvas over a thick padding. The canvas had holes lined with metal rings, which could be attached to hooks welded to the frames six inches apart. The ring ropes were already attached to the inside ring posts, which were connected to the corner posts by turnbuckles. Steel stools for the fighters were sitting outside two corners. They were attached to the posts in the corners by cables, thin enough so that they would not be seen on the film when the ring was sunken in the river. Marty and Jack attached the canvas to the frame and then hung gloves and towels on the ropes next to the corners where the stools were sitting. Towels were already laid in front of the stools. Gee, Marty had thought of everything. But how would we be able to move that contraption to our underwater stage? That was my job. How could I get it done? I was planning to carry it out in my canoe, but that was going to be impossible. I'd have to get Willie, the captain of one of the glass-bottom boats, to carry it for me. My only problem, then, would be how to get it down to the loading dock.

Six folding chairs were set up facing the ring. Jeb and

Woody were sitting in two of them.

When Marty and Jack finished their last-minute preparations, Marty opened a satchel and handed Jeb and Woody each a pair of trunks. He asked "Would you try these on?"

As Jeb and Woody ran out, Marty handed me the striped shirt and asked me to try it on. It fit perfectly.

When Jeb and Woody came back, we sat down in front of Marty, who continued, "This will be a lot more complicated than the 'shoot-em up.' It'll take a lot more time and will have to be shot in parts so that you can have time to catch your breath.

"Jeb and Woody will be the boxers. Newt will be the referee. He's the biggest. Sound effects will be added after the film is shot.

"For now, we'll run through the skit just like it's planned. Jeb and Woody, take your places in your corners. Reach up and grab your gloves and start trying to put them on. You'll need some help from Newt. He'll help Jeb first and then Woody. The gloves are made especially big so they'll slip on easily.

"When we start, Newt will be standing in a neutral corner, and he'll come over to help you, Jeb. You may not have time to get your gloves on before you have to come up for air, so just shove your left hand in a glove. If there's enough time, Newt can tie it for you. Then jump to the surface. I hope all three of you can come up and go down at the same time. After a few breaths, you'll go back down and take your seats again and continue putting on the gloves. Try to sit in the same position you sat in before you came up. We'll time you and have you come up about every two minutes. We'll have a cord hooked to Newt who can motion for each of you to come up. Take a few breaths and then go back to your posts.

"After Jeb has his gloves on, Newt will go over to help Woody, first with the left glove, then with the right. Again, you'll have to come up every two minutes. Please try to come up at the same time.

"When the gloves are on, you'll sit on the stools. Then I'll pull the cord and Newt will motion for you to come to the middle of the ring and tap your gloves. When you get up to go to the middle of the ring, you should kick your stools out of the ring, but don't kick too hard. We don't want to lose those things. At that point, you'll probably have to come up for air.

"When you go down the next time, you should go to the middle of the ring, touch your gloves once, then back away and come together boxing. Each of you will begin with a series of short jabs with the left hand. After several jabs, Jeb will make a right-handed hook that Woody will ward off with his left arm. You should repeat the sequence two or three times, but in the second series, Woody will make the right hook. After two series, come up for air. When you go back down each time, you need to try to return to the place you were before you came up. That will be particularly important for Newt. Throughout the jabbing, and after the hooks, Newt may have to separate you if you get into a clinch. Surely, after the third set of jabs, Newt will get the signal to push you apart, and you should go to your corners before coming up for air.

"The second round should be basically a repeat of the first round. But in the third round, I want Woody's hook to look like it's hard, and I want Jeb to fall to the floor. Newt will count him out.

"Any questions?" Marty asked.

There was a long pause while everyone started looking at everybody else. Heads were just turning from side to side. Marty gave such long-winded instructions that it seemed a real wonder everybody understood what to do. After the pause, Marty commented again. "Now let's have a dry run. Each man should sit in his corner. You'll need to try on the gloves this afternoon to be sure they fit."

I was impressed by how thoroughly Marty had worked out the details for a show that he could have only dreamed about.

I walked around the ring, climbed through the ropes and walked to one of the neutral corners. I stood there while Jeb and Woody climbed through the ropes and hopped into their corners to sit on their stools. They fiddled with their gloves until they were able to get them on. All I had to do was tie them on, one at the time.

Then Marty tugged on the line and I motioned for the men to go toward the center of the ring. Each man stood up and slid the stool off to the side of the ring, then walked toward the center. When I felt the line tighten, I motioned for the men to touch their gloves. Then I motioned for them to back away and come in jabbing. They followed instructions well but had to stop at that point to look as if they were coming up for air. When they acted like they were going back down, Jeb began his jabs, and Woody was not far behind. After half a dozen jabs, at least that's what it seemed like to me, they clinched. It almost looked to me like a love hug. I rushed in to separate them. Then they began jabbing again, and Jeb gave him a hook, but Woody warded it off. They backed away. Again, they clinched, and I rushed in to separate them. It was just at that point that the bell rang, and they ran back to their corners to sit on their stools before coming up for air again .

The second round was just like the first, but then came the third round, in which Woody threw a big hook and Jeb fell to the floor. I counted him out quickly and the fight was over.

We ran through the process again just to be sure we remembered it and then started getting the boxing ring and the rest of the equipment ready to take to the glass-bottom boat. I talked Marty and Jack into taking the ring and the equipment apart while I ran out to find Willie. That was easy. He was over at the maintenance dock tying up his glass-bottom boat for the night.

"Willie! Willie! I need your help," I told him.

"What 'cha got?" he asked.

"We've got a boxing ring we need to take down the river to a spot where we are going to film a boxing match underwater. I thought I'd be able to take it down on my canoe, but it's too big. It's a lot too big to take in my canoe. I need you to carry it down in your glass-bottom boat in the morning."

"You reckon I can? Is it too big for us to carry it on the boat? Can we get it there?"

"For sure. Marty and Jack had it sent from New York. The floor is the biggest part, and it is in pieces that can be taken apart. In fact, it's being taken apart right now. We'll be able to tie the big pieces on the side of the boat, and the small pieces can be carried on the main deck."

"How you gonna get it down to the loading dock and when? You know, I'm supposed to be at the dock at nine thirty in the morning to carry passengers."

"I'll find a cart and have it there by seven thirty, no later than eight o'clock. We can take it downriver and you can be back long before nine thirty."

"Okay. I'll have the boat there in time."

So that was fixed. My next job was to find some sort of cart. I ran in the maintenance shop and found Mr. Ritchey still there.

"Hi, Mr. Ritchey! Do you have a hand cart I can use to carry some pieces of a boxing ring down to the loading dock?"

"Yea. There's one out back, but what are you gonna use that for?" I asked.

"Those guys working for Grantland Rice want to make a movie clip of an underwater boxing match. They brought a boxing ring all the way from New York and had it set up over in the Okena Hotel. We had a practice session this afternoon, and now they want to move it down-river and set it up on the floor to make the movie. It's a sort of gimmick, but my job is to get it down there. I've talked Willie into taking it down tomorrow morning if I can get it to the loading dock before seven thirty. I wanted to use the cart to carry the big pieces over there tonight."

"Sure. You can use it. Just bring it back when you get finished."

I ran behind the shop, found the hand cart and pushed it over to the Okena Hotel. By that time, it was almost five thirty, so with Marty and Jack, we loaded the solid parts of the boxing ring on the cart and pushed it over to the loading dock. It was after dark when we got there, so we just unloaded the pieces on the dock and left them for the night.

I was up before daylight the next morning. I ate an apple and an orange before throwing cold water on my face. Then I ran down to the jetty and dived into the river before there were even streaks of daylight across the sky. Still naked, I swam the half mile to the lifesaving tower where I left my bathing suit the night before. Once I was dressed, I swam up to the sand beach and ran over to the docks outside the maintenance shop, where I loosened the ties to the pontoon raft and poled it over to the loading docks. I tied it right next to Willie's glass-bottom boat. By then it was fully daylight and Marty and Jack as well as Jeb and Woody were all there. We were ready for the work we would be doing on the bottom of the river. Together, we loaded the slats used for the flooring on the boat and put the mats and other pieces aboard. With five of us working, that took only a few minutes. I tied the pontoon raft to the back of the glass-bottom boat and away they went. I ran over to the sand beach to get my canoe and poled it down to the spot where the glass-bottom boat was anchored. Then I anchored the pontoon craft and tied the canoe to it. The water was almost ten feet deep there, and the floor of the river slanted slightly down toward the center of the river. We figured this would give the camera the best possible angle, with the pontoon craft anchored a little more toward the center of the river. It was almost as though the cameraman could look down on top of the boxing ring.

 We dumped the slats overboard, and we jumped in the river to guide them down to the bottom. We had all of the equipment off the glass-bottom boat within a half hour and were able to send Willie back to the loading dock in plenty of time to make his first scheduled run. Then we spread out the slats and locked them in position. We had to work two to three minutes at the time, but we were able to assemble the flooring before ten o'clock. We still had to attach the posts with the ring ropes and then the canvas. Each of us grabbed a post from the canoe and swam to a corner of the platform to place the post upright perpendicular to the floor. Next, was the canvas flooring which took a lot more effort to spread out and hook in place. That had to be done a little bit at a time with two swimmers at each side of the carpet pulling it tightly. The next step was to tie the ropes in position. Each rope had to be connected to the posts by springs and pulled around to the next post until all the ropes surrounded the ring. Then there were the stools to be wired in place and the gloves and towels to be put on the ropes near the corners. Each of the stools was in place by noon. In order to keep pilots of glass-bottom boats from running into the scenery, we cut four green bamboo poles and tied red flags onto them. We then tied a pole to each of the posts in the rink. We left the pontoon raft anchored by the rink as another sign for the pilots of the glass-bottom boats. With everything done, we were ready for a rest.

 I poled the canoe back to the loading dock. By then we were all so tired we decided not to begin working on the skit until the next day. After canoeing the photographers and the actors to the sand beach, I poled the boat back to the loading dock and warned each of the pilots to steer clear of the area around the flag poles.

 Marty, Jack, Jeb and Woody and I agreed to meet on the sand beach at seven thirty the next morning to go back to the underwater boxing rink. For our first underwater practice, I wore my striped shirt and khaki pants. When I strolled down to the sand beach at seven fifteen. Jeb and Woody were already there, but we had to wait a few minutes for Marty and Jack. They all piled into the canoe as I pushed it off the sand. Away we went to the pontoon raft where I hitched the canoe.

We were getting near the stage when Marty said, "Let's try to run through the skit just like we plan to do it when we start filming."

"How do you want us dressed?" asked Jeb.

"Just like you're going to be for the filming. This will be a dress rehearsal." Marty answered. "You and Woody will wear your trunks. Newt will be dressed just as he is now. What I need to find out is whether there are any changes in our original plans that have to be made.

"Newt will carry the gloves down and hang them on the ropes near your posts before you dive in. Then he'll come back up, and all three of you will dive down at the same time. Jeb and Woody, go to your corners and start pulling on your gloves. Newt will come over to tie them, first for Jeb and then for Woody. You'll probably be sitting down at the time, and you'll surely have to come to the surface at least once, probably twice before you get your gloves on. Remember that you will all come to the surface at the same time. Then you must come up after the gloves are tied before going down for the first round.

"Stand by your stools until Newt motions to you. When he does, kick your stools out of the ring and go to the center of the ring and touch gloves. Then back away when he signals before you make the first jab. I guess you'll get about a minute into the round before you have to come up for air. That will mean you'll come up right after Jeb fires his first hook. After you get a few breaths, dive down to finish the first round, then go back to your corners. Newt will give you the signal to surface if he thinks there's a need. His signal to surface will be pointing up with his index finger. If that is at a time we haven't planned, we'll just have to decide where to go back to while you are on the surface."

"Yes, sir" was all the actors could say.

Each man felt like he had been fully drilled. The big question was whether he could do everything he was supposed to do before he had to take another breath.

As it turned out, there was a problem with the flow of the river across the ring. The fighters were nearly swept off their feet by the flow of water every time they stood up. They had to change positions so that their feet could be planted against the flow. Each boxer had to move his post one corner to the left. It was even necessary to move the pontoon and start over so that the cameraman could be sure that he got his best shots. The new position made it more difficult for me to stand, but I didn't have to spar like Jeb and Woody, so I could handle my position.

We dived down again, taking up our new places. Then one at the time, I helped each man get his gloves on. After each pair of gloves was tied, we came up for air. When we went back down this time, Jeb and Woody rose from their stools and waddled to the center of the ring and touched gloves. When I signaled, they backed off and both came in jabbing. It seemed like a real match. They got locked in a bear hug once and I moved in to separate them. Then Jeb threw the right hook and Woody faked a fall, but he caught himself and was about to go back to jabbing when I motioned for them to go up for air. After only a couple of breaths, they were ready to go at it again. They dived down and repeated their jabbing. This time, Woody threw a hook at Jeb who warded it off beautifully. It was looking more and more like a real boxing match. It looked like we were ready for the real thing. After the bell rang above the water and Marty pulled my string to signal the end of the first round, I could see no need for more practice other than for the "knockout" that was scheduled for the third round. When we came up at the end of the first round, I told Marty and Jack what I thought and they agreed.

We then started with the third round and repeated taking positions as we did for the first. Then when the boxers came out, they began their jabs and continued for almost a minute before Woody threw his right hook and Jeb fell to the canvas. I counted him out. The fight was over. Both Marty and Jack thought they could get good pictures in the next match if all went as well as it had in the "dry" run. By that time, it was almost noon, so we gathered the props and climbed in the canoe, which I poled back to the sand beach. We decided not to try the filming until the next day.

The next morning, I was up before daylight once again. After washing my face and eating some fruit, I ran down to the jetty and dived in. I swam against the flow of water in the river the half mile up to the lifesaving tower, where I put on my bathing suit and swam on to the sand beach. I crawled out and ran over to the dressing room, where I collected the boxing gloves and towels for the performers. I put on my striped shirt and some khaki pants along with my brogans although I really don't know why I put the brogans on. I guess it was habit.

By the time I got down to the canoe with our props, Marty and Jack were there, and Jeb and Woody came running up within seconds. They all climbed aboard when I pushed the canoe off the sand. We were on our way to be movie stars, at least amateur boxing stars, in the depths of Sparklin' Waters River.

When we got down to the pontoon raft, we found it in the same place we had left it the day before. Flags were flying from bamboo canes attached to the corner posts where we left had them. I pulled off my brogans, and Jeb and Woody peeled down to their trunks. We dived in and untied the cane poles and laid them in the canoe where they would be out of range of the camera.

I climbed aboard the pontoon raft and lowered the tank while Jack was loading his camera. He then climbed into the tank and put the camera in place while Jeb and Woody and I were getting ready to dive into our places in the ring. Everyone seemed to be ready. First, I dived down to check to be sure the stools were in place. Then I swam back to the surface to wait for Marty to tell the three of us to dive in

Marty went over the list of actions. "Now remember, Jeb and Woody, take your seats in your corners when I give the signal. Then go to the center of the ring and tap your gloves when Newt gives you the sign. The camera will be running. After you have tapped your gloves and backed off, Newt will give you the sign to come in sparring. Simply follow the plan we made yesterday. Take your jabs, and Jeb will try a right hook that Woody will ward off. It'll be after that when Newt signals you to come up for air. If you need to come up before then, go ahead, but let Newt know. That way, we can all be in about the same state of readiness. Any questions? Okay, get in your positions!"

From the pontoon raft, each of us swam to our spots in the assigned corners. Within seconds, Marty gave the signal and we all dived down. The camera started rolling and they came together really fighting. There were only four or five jabs before they got into a bear-hug and I separated them, but by then, we had to come up for air. When we dived down again, the boxers took their positions and continued with the match. There was another right hook, this one by Woody, which Jeb deflected nicely. They took a few more jabs before I gave the signal for them to come up for the end of Round One.

We had several minutes to rest during which Marty complimented us on our performances. Jack agreed and assured us that he was getting some good shots. There seemed to be very little to change before we went down to start Round Two. That went about the same as Round One. We had a rest period about halfway through.

Now, it was time for Round Three, the knockout! We again dived down to our posts. Then Jeb and Woody headed to the center to begin their jabs. I joined them just in time to separate them when they went into a bear hug. They backed off, and Woody swung with a hook which Jeb tried unsuccessfully to ward off. Then Woody appeared to catch Jeb on the jaw. Jeb floated down, and then I had my big moment, counting him out. I did that with enthusiasm, swinging my right arm and hand. But I had to do it quickly so that we could get back to the surface for more air.

This time, when we boarded the pontoon raft to get the comments of Marty and Jack, they were tickled pink. The question was whether we could depend on the filming to be satisfactory or whether we should repeat the performance just in case it didn't turn out like we had wanted. After thinking it over, we decided to do it again the next day before taking down all that apparatus. The next day, dark clouds gathered as we sauntered down to the sand beach. Just when we got there, a bolt of lightning flashed. That was followed by a clap of thunder. We didn't dare go out on the river with that bad weather and besides that, Jack thought the clouds made the light questionable. We just had to wait for another day.

After a repeat performance the next day, we took down all the scenery, the ropes, the poles, the carpet and the flooring and hung it on the side of the pontoon raft. I called Willie to cart it back to the loading dock where we dumped it. I got the cart back and we carried the metal ring back to the hotel. Then I loaded the canvas and the mat into my canoe and poled it down to Camp Nasty for drying.

Marty and Jack carried the film back to New York to be developed. We waited for the results, and it was a week before Jeb's phone rang.

"Hello," said Jeb.

"This is Marty. I just called to tell you that the film turned out better than we could have ever hoped."

"Great. How's the boss like it?" Jeb asked.

"Mr. Rice? Oh he's happy. At least he didn't have any catty remarks. That's always a good sign. I think he really liked it. My guess is he'll use it in one of his shows right soon," Marty said.

From all appearances, Grantland Rice was happy with it. In fact, he showed some brief clips as introductions and trailers on his news films that week. The film was the talk of the town.

11

ABOUT A MONTH after we finished the underwater boxing match, Jeb received a call from Marty telling him that Grantland Rice's crew wanted to come back to Sparklin' Waters to film some more stunts.

"What do you want to shoot this time?" Jeb asked.

"How about shooting a man smoking underwater?"

"What? How's that possible?" Jeb demanded to know.

"Well, I saw the gas swirling toward the surface when dry ice was held down in a glass of water. It looked just like smoke coming from a cigarette. I thought maybe we could whip up a skit using the dry ice rolled in paper and make it look like a cigarette. I'd also like to film another skit, an orchestra performing underwater."

"That sounds kind of kooky to me, but I'm game. We'll try anything," Jeb said.

"Do you think we could film the two skits next week?" Marty asked.

"I'll check and see if anything is scheduled," Jeb responded.

Jeb checked with Shorty and found that the week was free. He called Marty back. "We can work next week. I'll make some reservations at the Okena Hotel for you. When are you getting here and how long do you think you'll stay?" Jeb asked.

"We'll get there Sunday afternoon, and we'll plan on

being there the whole week. We can see if the cigarette stunt will work on Monday, but it may take several days to do the orchestra skit. We can meet at the hotel at nine o'clock Monday morning to talk about the skits. Can Newt have the pontoon raft ready Monday morning?" asked Marty.

"I'll see if Newt can be free and get the raft ready. Then I'll make the reservations and call you back. If all that can be worked out, I'll have the whole crew at the hotel at nine Monday morning. I'll probably be there ahead of time," said Jeb.

"By the way, could you check with a nearby ice cream plant and see if you can get five pounds of dry ice? You'll have to keep it in an icebox. It's extremely cold, so be careful. Holding it with your bare fingers will give you frostbite in just a few moments. If you bring the ice to the hotel when you come, that will save us a lot of time," Marty said.

"Fine. See you then," said Jeb.

On Monday morning, I climbed out of my bed as usual and, after a fruity breakfast, swam upstream to the lifesaving stand where I put on my bathing suit. Then I swam over to the sand beach and went to the dressing room, where I dressed in my usual khaki pants and slipover shirt then ran over to the dock where the tank was anchored. I untied the ropes and poled the apparatus over to the loading dock and tied it down. By that time, it was almost eight o'clock, so I looked around until I found Jeb, and we took a little ice chest over to the Mid-Florida Ice Cream plant where we bought a five-pound block of dry ice. The plant was about two blocks from the Okena Hotel, and we got there just as the town clock was striking nine. At the desk clerk's counter, we asked, "Are

Misters Marty Warren and Jack Eaton here?"

"Yes, sir. They checked in last night. They sure had a lot of luggage."

Just then I turned around to see both men strolling into the lobby from the dining room. They were grinning like Cheshire cats.

"Hi! Glad to see you made it," I said to Marty and Jack as they came over to the hotel desk where we were standing.

We shook hands. Then Marty turned to the desk clerk and asked, "Is one of the conference rooms free? We'd like to use it this morning."

"You can use No. 2. It's free all day."

We went over to the conference room and pulled the chairs into a circle. Marty sat down and began talking. "We've found a way to make a picture look just like somebody smoking a cigarette while underwater. We want to give it a try. We think it would add a lot to our skits. Let me have some dry ice. I'll cut off a piece to show you just what I mean."

Marty pulled a folded sheet of typing paper out of his pocket and cut off a strip about three inches square. Then he put on a pair of cotton gloves and opened the ice chest. He pulled out a big pocket knife and cut off a strip of the dry ice about four inches long and about a quarter inch square. He wrapped it in the square of paper, making it look like a cigarette, except that the ice was sticking out one end of the wrapper. I could see the gas escaping from the dry ice right there in the room. He then took a glass jar and held the make-believe cigarette underwater and it really did look like smoke pouring from a cigarette.

Marty then held the cigarette up to his lips, pretending like he was smoking it.

"It'll look a lot more realistic in the river," he commented.

"Jeb, Newt will have his hands full helping us get the tank in position. Would you put on some torn jeans and this derby hat and be the subject for this movie? By the way, I forgot to mention that there's nothing coming from this but carbon dioxide. So, there is no danger, except from holding the dry ice with your bare fingers, and that's not a problem when the rolled cigarette is held underwater."

Jeb was nodding agreement even before Marty got his sales pitch out of his mouth, but to make it clear, he said, "Sure. I'll dress for the occasion."

With that, everybody except Jeb headed for the door and walked down to the loading dock where the pontoon raft was tied. I ran over to the sand beach, pushed the canoe back off the sand, poled it over to the loading dock and tied it to the raft. By the time I got there, Jeb had gotten dressed in his old jeans and his derby hat. He looked like a real bum. But with our morning planned, we had no time for small talk. It was down the river for almost half a mile. We were very near the place where we had taken the pictures three weeks before.

I anchored the pontoon raft and put some water in the tank so it would sink. When the tank was resting on the river floor, I tightened the bolts and emptied out the water.

Jack climbed into the tank with his camera and called out, "I'm ready, Jeb. You can jump in front of the screen."

Marty began directing once again. He handed Jeb the make-believe cigarette and told him, "Hold this in your gloved hand until you get down. Then hold your hat in one hand and the cigarette in the other until you are in front of the tank. Next put your hat on, turn your side to the screen and put the cigarette up to your mouth like you are smoking it. The gas will take care of itself. Turn in several different positions so we can get shots from each. After a couple of minutes, come on back up and we'll see what we have."

I was standing beside Marty and watching from the frame connecting the pontoons, but the whole scene looked real to me. I didn't know what it might look like in front of the camera.

When Marty came up, we congratulated him. Jack came out of the tank and congratulated him as well. He said, "That was perfect. We might do it just one more time to be certain that we have a good film."

After a few moments, Jack climbed back in the tank, and Jeb jumped back in the river and went through the skit. Again the scene looked realistic from where I was standing. Jack was satisfied as well and Marty agreed, so we finished our work for the morning. I poled the pontoon raft back to the loading dock and lashed it down. After it was emptied, I poled it back to the maintenance dock to be left for the night. Then I poled the canoe back to the sand beach.

"PLEASE. CAN SOMEBODY help me lift this case? We've got to carry it into the conference room," Jack said.

Jack was pushing a hand cart loaded with one huge leather case and several smaller cases up to the door of the conference room but could not get them through the door.

I jumped at the chance to help Jack. Sizing up the situation, I saw that the problem was that the biggest case was too big to get through the door unless it was turned lengthwise. So I grabbed the handle to get some idea of how heavy it was. To my surprise, it was quite light, so I simply lifted it off the cart and carried it into the room. The other cases were easier to get through the door. In fact, it was possible to push the cart, still loaded, through the door, but it was obvious that it would take up a lot of space if we kept it inside, so I just unloaded the smaller cases, carried them in and pushed the cart against the wall in the hall.

Once the cases were set on the floor in the room, Jack started opening them and discovered that each case held a musical instrument. We found a tuba, a French horn, a trumpet, a saxophone, and a clarinet.

We took each of the instruments out, assembled all that were broken down and laid them on the table in the front of the room. Just then Marty and Jeb walked in. Martha, Eileen and Elsie were right behind them. Each one picked up one of the smaller instruments, held it up to her lips and blew. I was surprised to see that Martha could make good music with the saxophone. Eileen picked up the trumpet and blew into it enough to make a few sounds, but she was a long way from being able to play a tune, at least any tune I had ever heard, maybe except from the outhouse.

When Marty walked in, we laid the instruments on the desk and took our seats. In all, there were three men and three women from Okena and the two fellas from New York.

"This may be the most difficult skit we have tackled," Marty commenced. "I want to get a clip of you playing in a band, creating the image of an orchestra. Newt, since you are the biggest, we'll let you play the tuba. Jeb, you can play the French horn, and Woody, I'd like you to play the trumpet. Martha, you have already shown that you have lots of know-how on the saxophone, so we'll put you in charge of that. You already know how to hold it. You may be able to help Eileen with the clarinet. Elsie, we have a flute for you.

"Now, grab your instruments and play around with them so you can get used to fingering them or at least holding them. Newt, yours will be the most difficult to manage. The big bell that attaches to the main part of the tuba may be caught up by the current in the river. You'll have to be certain that you always have the bell turned crosswise so it won't get caught in the current. All the other instruments should be easy to manage. This afternoon, what you need to do is to get to know your instruments and get some idea of how to hold them to make it look like you are playing them.

"One other thing—everybody has to have a stool to sit on. I've borrowed some from the hotel. They'll have to be wired so that the wires won't be seen on the film, but when we get down to the river, don't let the stools get away and be washed downstream."

Everybody grabbed an instrument and sat back in his or her chair. I held my tuba up so that the ring of the instrument curled around my body. I even tried to blow into it and it made an awful noise, like I was passing gas. I was. Martha played a little tune on her sax. She was the only one able to make a recognizable tune even if it was "Mary Had a Little Lamb."

We sat back in our chairs, and Marty came around to help us get our instruments in position. While he was doing that, Jack brought in six steel bar stools. We pushed the chairs to the back and tried to arrange them so that

Jack would be able to get each of us in the pictures being made of the whole group. After that, we tried sitting on the stools until we felt comfortable. Everything had to be arranged to get the setting Marty wanted. Getting positioned took a lot more time than we expected. It must have taken us at least an hour.

We were trying to plan everything ahead of time. Still, we had no idea how long it would take us to get set up in the river and how many times we would have to come up for air. Obviously, I was going to have the hardest job because of the size of the tuba.

After we had fiddled around long enough to get an idea about how to hold our instruments and what our relative seating arrangements would be, Marty spoke up.

"Can everyone be down at the loading dock by nine o'clock tomorrow morning? I'd like for the men to meet at the hotel at seven thirty and help load the instruments on a cart to take them to the loading dock."

The next morning, when we got the instruments to the loading dock, I ran over to the sand beach to get the canoe and pontoon raft. I tied the canoe to the frame connecting the pontoons and poled the contraption to the pier, where we loaded all the equipment, including the stools and the instruments. We had to be sure there was some wire to hold the stools when they were on the river bottom.

The girls climbed into the canoe with Jack, and the tuba and French horn were hung on the frame of the pontoons. The smaller instruments were in the canoe that was left tied to the pontoon raft. I poled the whole gizmo downriver. When we got to the place where we planned to make the pictures, I anchored the pontoon raft so that the window of the tank was facing the bank of the river.

We then tied the stools together and dumped them in, but I had to dive down and bury the legs in the sand so they would stand up. Marty kept watch to see that they were placed in the right position so that the camera could catch them like it did at the hotel.

We actors took off all our clothes except for our bathing suits and grabbed our respective instruments. We jumped in the river and scrambled onto our stools. Then we raised our instruments as though we were preparing to play and held them in place for a minute before we had to come up for air.

The trial showed us how poorly we were prepared for the water current and weightlessness. Jeb was holding the French horn upside down, and he couldn't find the keys. Martha still had her hands on the saxophone, but it was all she could do to get the mouthpiece between her lips. Eileen held the clarinet straight out in front of her, but that wasn't so bad, and Woody had to search around to get his lips on the mouthpiece of the trumpet. Elsie seemed to find the natural position for the flute, but we had to get her stool in position so the flute could be seen on the film. The bell of my tuba was caught in the current in spite of all the precautions. It took me a while to get it turned so I would not be carried away. We fiddled around with our instruments only a minute before we had to come up for air. When we came up, everyone brought up his or her instrument except for me, so we had to start all over again with the positioning, but each time we restarted, it was easier than it was the time before. Of course, it was harder for me because the bell of that tuba was so big.

When we got on top of the water, we were able to explain our difficulties to each other. We knew that it was the river that was giving us so much trouble, and we knew that it was just going to take time for us to get used to the instruments underwater before we could think about getting movie shots.

Marty said, "Do you think that it's too much of a job, or do you think you need to just spend more time getting used to having the instruments underwater?"

"Oh, I'm sure we can get used to them and all this will be lots of fun. It's just going to take time. Let's try to go down a couple more times this afternoon and then wait to start filming until tomorrow," I suggested.

"Sounds great to me," Marty replied. He was obviously relieved.

"We'll give it a try. We can always scrap the plan if we aren't able to do it."

We were determined to be successful. The next dive into the river went better than the first. At least everyone was able to find his or her seat, and we all found good positions for our instruments.

On the third try, I was able to hold the tuba in a position so that it would not hide the pictures of the other instruments but would also still be in the pictures. Martha's saxophone looked natural and Jeb was able to keep track of the mouthpiece of his French horn.

By that time, we all seemed to be getting more used to our positions, and we felt we could hold them while the films were being shot, so we quit for the day with a plan to return at nine the next morning.

I was awake at six o'clock that next day. I grabbed an orange and an apple for my breakfast. I ate the apple first, then peeled the orange and gobbled that down on my way down to the pier. I dived into the river and swam up to the life-saving tower to get my bathing suit. After pulling it on, I swam over to the sand beach and ran on to the maintenance dock, where I untied the pontoon raft and poled it over to the loading dock to wait for everybody else. In the meantime, I ran back to the sand beach and poled the canoe over to the loading dock so everyone would have a place to sit until we got out to the set. By then, it was nearly eight, so I had over an hour to wait. As it turned out Marty and Jack got there early. Jack was weighed down with his camera and film, so I helped him get both in the canoe. They waited on the pier until Jeb, Woody, Martha, Eileen and Elsie arrived just before nine o'clock. They all sat in the canoe while I poled downstream until we got to the place where we had planned our underwater platform. I anchored the pontoon raft with its tank in a position so that the window of the tank was facing the underwater musical stage.

    Everybody peeled down to a bathing suit. We grabbed our instruments and were ready to jump in, but then we decided it might be better to dive down to see where our stools had wandered and then come back up to get our instruments. We hopped in and repositioned the stools that had been tilted by the water current through the night. The legs of every one of them were washed out of the sand. Then we straightened them and reburied the legs. We swam back to the surface and to the canoe to get our instruments.

    After catching our breath, we jumped back in and took our positions with our instruments. Marty was watching all of this and motioned to Jack to begin filming as soon as we were perched on the floor of the river. There was not much movement to be taken, but we tried to look like we were actually playing instruments.

    After a couple of minutes, we came up for air. Marty

was happy with the positions, and he thought the pictures would turn out well, but he asked us to repeat it twice to be sure that they had enough good film exposed. After it was all over, we came back up and left our instruments in the canoe. We then went back down to get the stools and anything else we might have left. We poled back to the loading dock, where I left the other members of the crew, and took the pontoon raft back to the maintenance dock to be tied down to wait for the next movie production.

12

"**NEWT, I TALKED** to Shorty Mayson today about the possibility of making a training film on how to prevent pain in the ears and nose while swimming and diving. He told me to get together with you. He said you manage the tank used to shoot the pictures, and he said that you are a really good swimmer. He thought you might help with the tank and maybe you could help me get set up to make the films I want."

"I'll sure give it a try," I said.

That was my introduction to Dr. Marshall Taylor—or perhaps better said—his introduction to me. Dr. Taylor found me at the sand beach. It was late in the afternoon and I was nearly finished with my day's work. We walked over to Dr. Taylor's car, from which he took out a big suitcase. Then we walked over to the Sparklin' Waters Grille and sat down at a table. Dr. Taylor ordered Cokes, and we sat at a table near the front window, where he started to explain what he wanted.

"Newt, I want to make a film to show what causes people to have pain in the head and even earaches sometimes when they go swimming. I have a model of the upper part of the body with the air passages. It's made out of plastic so that a person can see inside and, with a little guidance, can understand what I'm trying to show. What I need for you to do, besides moving the tank around for me, is to put some probes up through the nose and in the right places inside the nose and back in the throat. Here. Let me show you," said Dr. Taylor.

With that he pushed his Coke aside and reached down for the suitcase. When he opened it, he pulled out a life-sized, upper-body dummy with a glass head and shoulders. The naked bust showed everything down to the belly button.

When Dr. Taylor laid the bust on the table, he pulled out some little steel probes with handles about six or eight inches long. He took one and handed me one.

"First, let me show you the spots where I want these probes placed," he said.

With that, he pushed his probe up the nose to a spot under the big fold in the upper part of the nose. He called that the upper turbinate. I had never seen that before. Then he pushed the tip into a little hole there. After he showed me what he had done and explained it to me, he then moved the probe to the back of the nose, where the tube from the ear joined the throat, and he left it there. He called the tube "the eustachian tube" and that part of the throat the "nasopharynx." Then he reached down in his suitcase and came up with another probe, this one with an angle on it, and he pushed that down between the vocal cords.

"These probes will have to be put in one at a time," he explained.

"What I want you to do is take the bust and dive down to the bottom of the river. Set it up on the sandy bottom. You'll have to bury it a little way and pile sand around so it won't be swept downriver.

"After that, I want you to place the first probe in the hole coming into the nose from the sinus and let me get some shots. Then I'll let you put it in the place where the eustachian tube enters the nasopharynx, and then later, I'll want you to put one in between the vocal cords. These will all be done, one at a time, and, each time, I'll take some pictures when you come up for air."

After his little training session, I thought I could handle his probes, so I felt comfortable about the work. We agreed to meet the next Saturday afternoon after the tours were finished at the loading dock where the glass-bottom boats were tied up.

When Dr. Taylor arrived, I had the pontoon raft at the docks. While still at the pier, I showed him how to climb into the tank. Then I helped him get in the middle of the canoe. I left the canoe tied to the pontoon raft, and I poled down the river a few hundred yards. When I got to a quiet spot, I anchored the pontoon raft and had Dr. Taylor climb aboard the raft with his sixteen-millimeter camera. I put the raft in a good position so that the window of the tank would be facing the incline on the edge of the river. I then flooded the tank with water until it rested on the bottom of the sandy riverbed. Next I clamped it down, dipped out the water and let Dr. Taylor climb inside with his camera. Once Dr. Taylor was inside the tank, I covered it and stuffed the pointers inside my bathing suit. Then I held the dummy under my left arm and dived into the river. Dr. Taylor stood in the tank and watched me while I found a good spot to secure the dummy. Then I had to put some sand around it so that the current of the river would not pull it loose, and I had to come up for air. When I went back down, even before I put the probe in the nose, Dr. Taylor started shooting pictures. When I put the probe up the dummy's nose, I was able to put it right in the hole between the turbinates. Then I came up to breathe.

I was hardly at the surface when Dr. Taylor said he was ready for the pointer to be moved to the back to the nasopharynx. I dived in and moved the pointer, and again, he started shooting while I was moving the probe. After that series of shots, Dr. Taylor pulled out an anatomy book and showed me how to get the pointer between the vocal cords. When I put it there, the tip of the probe was sticking into the windpipe. Each time I moved the probe, he wanted some adjustments, which I was happy to make. He also asked for some other placements, which took all afternoon—that is, we kept on until it was getting dark, too dark to take more pictures. By that time, Dr. Taylor had all the pictures he wanted anyway. The whole process took three or four hours. I then had to get the dummy and bring it back to the canoe. After that, I poled the canoe and pontoon raft back to the pier.

While we were going back to the pier, Dr. Taylor told me that he wanted me to come to his office in Jacksonville the next time I was over there. When I went, he took me inside and made me take a seat. Before I knew what was happening, he was checking my eyesight. I often wondered whether he thought I couldn't see what I was doing down on the bottom of the river. At any rate, Dr. Taylor appreciated my help with his dummy so much that he gave me free eye care for as long as he lived. I couldn't have appreciated that more.

13

"HAVE YOU BEEN to the picture show lately?" Cousin Jeb asked me one afternoon. That was several months after I started working at Sparklin' Waters.

"Naw. I don't go to picture shows very much," I replied.

"You ought to take the night off and go downtown with me tonight to see the first showing of *Tarzan, the Ape Man*. You'd like that. It's a good movie. Johnny Weissmuller is the star. You know, Johnny was a champion Olympic swimmer. What's more, a major part of the movie was shot right here in the Okena National Forest. It shows a lot of the swamp and jungle with all the tangled underbrush. And what'll dazzle you is that Johnny is pictured swinging through the trees on the vines. You won't get a chance to see that very often. There is a trek through the hills and even a shipwreck," Jeb said excitedly.

Before Jeb talked me into going with him to see that movie, I hadn't thought much about it. I had heard that a movie had been shot around Sparklin' Waters, but I never dreamed that the show might have anything of interest to me. Even after I saw the picture, I never dreamed that I might get to know Johnny Weissmuller, much less become friends with him.

After Jeb talked me into it, I took off that night and

went into town with him to see the show. It was true, I hadn't been to any movies since I had left Ashburn and really didn't know what to expect. When we got downtown, there they were milling around in front of the box office: Johnny Weissmuller, Maureen O'Sullivan, and the whole lot. Before we saw the picture, I couldn't imagine anyone swinging through the jungles out there like Tarzan did. I never tried it and really didn't dare, that is, unless I got paid for it, even after I saw him do it. But one other thing caught my attention. All Johnny wore was a loincloth, a sort of G-string, and Maureen O'Sullivan just a wore little more. I didn't know they made movies like that, but then everything seemed so real, so natural, out there in the jungle. I really liked the show.

As for the setting, I had a hard time placing it. According to the story, it was supposed to be in Africa, but I knew from what Jeb had told me that most of the movie was shot in the United States, most right there in the Okena National Forest. Since I had gone hunting out in the Okena National Forest several times with Rob Allen, the jungle scenes made me feel right at home except for the mountains and wild animals. In the movie, there were lots of zebras, lions, tigers, and hippos as well as monkeys and apes. In fact, Cheetah, a female ape, was a star of the show.

When I first went to work at Sparklin' Waters, I saw rhesus monkeys scrambling through the woods. I was told that two pairs of them had been brought to Sparklin' Waters years before by Col. Tooey when he was putting together a sight-seeing tour up and down the river. Col. Tooey owned a fleet of sight-seeing boats up in New York, where he held tours during the summer, but at the end of the season, he moved his touring business to

Sparklin' Waters. He started by towing a couple of speed boats down to the resort to set up jungle cruises. At first, he sold a good many tickets, but there were not many tropical animals for his curiosity seekers to look at, so the popularity of his cruises dwindled. Col. Tooey bought the rhesus monkeys and let them run free in the swamps. He built a shelf along the side of the river and stocked it with apples, grapes, bananas, and other tropical fruit just before a cruiser was scheduled to pass under it. That way, the monkeys would come to the riverbank to be seen by his passengers. In time, the monkeys reproduced, but after a couple of generations, the young ones began to look sort of sickly. They came down with all sorts of diseases. As time went on, many historians tried to blame the movie companies for bringing the monkeys to the area, but the monkeys were there long before the movie companies. Indeed the ones that they brought to use in the movies came from zoos, and they were usually tame. After the movies were shot, they were taken back to their stateside homes. More recently, the health department has tried to get rid of the monkeys that still have not adapted to the wilds of the Okena Forest.

Ultimately, regarding the scenes in *Tarzan, the Ape Man,* I never saw any mountains in the jungles where Rob took me during our hunting trips, and I didn't remember seeing even any big rocks shown in the movie. I also never saw any of those wild animals pictured running across deserts, and I knew there weren't any deserts in the Okena National Forest—at least, I had never seen or even heard about any.

There were also some big snakes in the movie—pythons or boa constrictors. I never saw any of those snakes, and I knew that if Rob Allen had ever seen any, he would have told me and he would have certainly been trying to catch some. He and I caught moccasins out in the swamps, and he stored all the snakes he could find in the pen he had built in front of his house.

While we were walking back to Jeb's house after that picture show, I asked him about those scenes, and he told me that there were several scenes shot in places other than the Okena National Forest. He thought that most of them were made in different places in California, but he wasn't sure. The movie makers certainly did a terrific job of blending scenes from different places together. I couldn't tell where some of the scenes began and others ended. I guess that's because the producers were so good.

As it turned out, it seemed as if everyone I talked to really liked the movie, and, even though I didn't read the newspapers, I heard that the movie reviews were good all over the country. People's only negative comments had to do with Tarzan's acting when he was not swinging through the trees. I heard a few weeks later that MGM was thinking of making at least one other movie about Tarzan.

In the meantime, I grabbed every chance I had to learn all I could about Edgar Rice Burroughs, the author of the book on which *The Ape Man* was based. I wanted to know where he was born and grew up and how he came up with such a fantastic story.

I found some articles that said Burroughs was born in 1875 in Chicago. His family was pretty much normal except that they had money. The boy had a normal life until he was a teenager, at which time schools in the Chicago area closed down because of the flu. His folks sent him to live on a cattle ranch in Idaho owned by his two older brothers. He rode the range, herded cows and broke in a wild horse. He really liked the wild west, but he got mixed up with some cattle rustlers and a bunch of other men with bad reputations. When his folks heard about that activity, they hustled him off the range. They signed him up for Phillips Academy in Andover, Mass., but he wasn't there long before the faculty decided that he just didn't fit in with their other students. At least he didn't act like them. He seemed to get into lots of trouble. The family then sent him to Michigan Military Academy, where his western lifestyle seemed to fit in better. He even did pretty well in his schoolwork.

By the time Burroughs graduated, he decided that Army life was for him. He got an appointment as an instructor at the academy and applied for admission into West Point. But he failed the entrance examination, so he quit his job at Michigan Military Academy and enlisted as a private in the Seventh Cavalry in the Arizona Territory. He still had hopes of becoming an Army officer.

Burroughs spent a lot of time and effort chasing Apaches in Arizona, but he never caught any. He did catch the diarrhea. When he was examined by a doctor, a heart murmur was discovered. That ruled out any possibility of his ever getting a commission in the Army. He was discharged from the Army in 1897 and went back to Idaho to ride the range. He even ran a dry goods store for a short time.

After staying in Idaho for two more years, Burroughs moved back to Chicago to work in his father's battery company. There he married his childhood sweetheart.

In 1904, Burroughs took his bride back to live in Idaho. Then later Edgar took a job as a policeman in Salt Lake City but gave that up to go back to Chicago. There, he had several brief jobs including door-to-door salesman; accountant; manager of a clerical department of Sears, Roebuck and Company; peddler of a quack alcoholism cure; and finally, a pencil sharpener wholesaler. By that time, his wife had borne two children and he was flat broke. His greatest peace of mind came when he was escaping the real world by daydreaming. He must have had some terrific dreams.

To sell pencil sharpeners, Burroughs had to advertise. Looking for the best places to put his ads, he began thumbing through some pulp fiction magazines. Since his salesmen were not bringing in much business, he had a lot of time to read short stories. He was cocky enough to think that he could write rubbish at least as good as that which he was reading. Since his family responsibilities were growing, he decided to give writing fiction a try, even though he had no training or experience in the field. He wrote the first half of a novel he named *Dejah Thoris, Princess of Mars*. When he sent it to a publisher, the editor liked what he read and told him that if the rest of the novel were as good as the first part, he'd buy it. When Burroughs sent in the second half of the manuscript, the publisher changed the name of the book to *Under the Moons of Mars* and paid him four hundred dollars. When he looked back years later, he reckoned that no amount of money would ever give him the thrill that he had received from that first fee. He decided then and there to make writing fiction a career. However, he still kept a salaried job, at least for a time. Eventually, he became a department manager for a business magazine.

It was while Burroughs worked as a department manager that he wrote *Tarzan of the Apes*. He wrote in longhand in the evenings and on weekends and holidays. That novel was published in a magazine in 1912. For it, he received seven hundred dollars. That same year, he wrote *The Gods of Mars*, which sold immediately.

Burroughs then went back to writing about Tarzan. He wrote *The Return of Tarzan* during the months of December 1912 and January 1913. When he sent it in to his original publisher, it was rejected, but another publisher paid him one thousand dollars for it.

By this time, Burroughs' third child was born, and he decided to give up his salaried job and spend all his time writing. He then had no steady income. Having sold his manuscripts for publication in magazines, he was not getting any royalties. But every well-known book publisher he could find turned down his tales of Tarzan. If he had not been able to sell his writings to magazines, neither he nor his family would have been able to eat. Fortunately, he was able to sell his stories for cash. Eventually—and this was long after I started working at Sparklin' Waters—he wrote twenty-five novels about Tarzan. Twelve movies featured Johnny Weissmuller as Tarzan. Through the years, fifty-three movies about Tarzan were filmed. Metro-Goldwyn-Mayer (MGM) made most if not all of them.

The owners of Sparklin' Waters wanted to get as many movies as possible made in the river and surrounding swamp and jungle area. They thought that was good advertisement. To entice the movie companies and yet keep control of the activities, the owners offered not only the facilities to help with the filming, but also any employees who might help them. Workers at Sparklin' Waters were told to take off when they were wanted by the filming companies but to return to their assigned duties when they were no longer needed to help with the movies. When they were helping, they were paid by both the resort and the film producers. Employees couldn't be happier. The only demand the company made to the movie producers was that they make notations that the facilities of Sparklin' Waters were used. Those notes were to be shown at the start and end of the show, whether a short clip or a full-length movie.

As time went on, those of us who worked for Sparklin' Waters became more useful to the movie companies. We worked closely with both cast members and producers of the films. My primary responsibility when I first went to work at Sparklin' Waters was managing the tank for shooting underwater movies. I remember especially well that when they were making *Tarzan and His Mate*, Tarzan tore off Jane's clothes just before she dived into the river and swam naked. The movie company ordered all the personnel not needed for the filming to leave the area. Since I operated the tank, I was able to watch those scenes, but my view was not all that good. What I did see was really exciting, especially for a young man just in his prime. As a matter of fact, when I saw the finished film, I thought it was fantastic. I couldn't see anything wrong with it. The naked swimming scene was one that I will always remember. Nude scenes in movies were later outlawed in the United States, but there are still copies of *Tarzan and His Mate* made in black and white.

Those of us who worked at Sparklin' Waters came to play bigger and bigger roles in the productions. A lot of the locals played parts in the filming of the Tarzan movies, and those experiences surely had an effect on the future careers of some of those people. We locals were usually used as "stand-ins." Goolie Green stood in for James McArthur who later played Dano in *Hawaii Five-0*. Elsie Nixon stood in for Maureen O'Sullivan on occasion, and she was the one filmed in the nude swimming scenes. Rob Allen and Eileen Perry-Waddington each played bit parts.

Most of my acting experience came about during the filming of *Tarzan Finds a Son*. I was used as a stand-in for Henry Wilcoxen to lead the safari. I still have pictures taken after those scenes were filmed. Almost one hundred black men were used in making that picture. They carried boxes or bags of supplies bound with hemp into and out of the jungle on their shoulders or on their heads. That was the way the producers got equipment to the scene. These same men were used as extras during the filming. Most were dressed only with G-strings, and many had their faces and upper bodies painted to make them look like savages.

Before making that film, I had never realized just how much imagination Burroughs had. In the movie, a baby boy was lying in a crib on a flight with his family to South Africa in a prewar Lockheed airplane that developed engine trouble and crash-landed in the jungle. Everyone on the plane except the baby was killed. After the plane crash, the baby's crying attracted an ape, which rescued him. The ape gave the baby to Tarzan, who took him to his primitive home made of bamboo with a thatched roof, built high on a rocky mount. When Tarzan's wife, Jane, saw the baby, she set about taking care of him, temporarily at first, but when they found the wrecked plane with no one alive, their newfound baby became a permanent part of their family. When Jane tried to select a name for their new family member, Tarzan demanded that he be called Boy, and that name stuck. As Boy grew older, Tarzan taught him how to survive in the jungle.

It was five years later when a search party, some relatives and some white explorers came looking for the plane and any possible survivors. When they discovered Boy and found the wrecked plane empty, they tried to take Boy back to England, but Tarzan didn't want to let him go. Boy had come to love his wilderness and didn't want to leave. By this time, he had learned to swing through the trees and give out calls, just like Tarzan. Jane was torn between sending Boy back to England where he would be educated for a place in a modern, sophisticated society and keeping him in the only home he had ever known. In the end, he remained in the wild with his jungle family.

While Boy's relatives were visiting the jungle, Tarzan played host. He swung through the trees and ran through the jungle fighting off attacks by wild animals. He had to rescue the explorers from not only the wild animals but also savages. In his travels through the jungles, Tarzan frequently rode the backs of huge elephants and Boy rode behind on a baby elephant. One of my joys during those experiences was watching the elephants play with the crew. During the filming of *Tarzan Finds a Son*, a baby elephant rolled Boy around on the ground.

Those were the days! Gee, what an imagination Burroughs had!

As for me, I had many personal experiences with Johnny Weissmuller during the filmings. We came to be good friends. Later, when I began driving long-distance buses and Johnny was living in South Florida, he would come out to meet me at the bus stop in Fort Lauderdale.

I also made a number of other friends in the movie industry because filming the Tarzan movies brought a lot of Hollywood personalities to Okena and Sparklin' Waters. Famous performers and their families usually stayed at the Blufton Hotel in Okena. In addition to screen idols Johnny Weissmuller, Maureen O'Sullivan, John Sheffield, Ian Hunter, Henry Stephenson and Frieda Inescort, the director and producer of the movies came to Okena for a time. Beryl Scott, Johnny Weissmuller's wife who was a beautiful blonde, also brought their two boys to town every few weeks.

I was even offered the opportunity of moving out to California to work in other movies, but I decided that this side of the country was where my bread was buttered.

14

"NEWT, GIVE'EM A tour."

That was my order from Captain Shorty when Rob Allen took charge of the Seminole Indian reservation. Shorty was trying to develop the Indian reservation into an added attraction for Sparklin' Waters. A special fee was charged for visiting the Indian sanctuary.

Rob Allen was a short man, almost 37 years old when I first met him. He was full of life, always trying to figure out how to catch wild animals. He was especially attracted to snakes and alligators but would collect most any kind of wild animals he could find around Okena. He was born in Pennsylvania in the early 1900s. He grew up and married a Quaker when he was still young. He and his northern wife had two children but they just didn't get along together, so he got a divorce and moved to Sparklin' Waters.

When Shorty ordered me to give our visitors tours through the Seminole reservation, I started visiting the Indians whenever I could and I tried to learn all I could about them.

I learned a lot about the Indians' habits. They cooked in a single pot over an open fireplace sunk in the ground. Fresh hams were hung from racks where they mellowed before they were cooked. The meat was covered with sprouting hearts of sago or cabbage palms while it was being cooked. Hearts of palm were hacked from the tops of trees. The Indians called the hearts of palm "swamp cabbage" or "cabbage palmetto." I always cooked the palm along with french fries. Only the part that could be broken by hand was cooked. If you had to use a knife to cut the palm, it was too stringy and would get stuck in your throat. I just have to tell you, you can eat your belly full of palms, but the next day, you won't need any medicine. It really cleans you out.

I never ate with the Indians. I thought their habits were just too nasty. One thing that really bothered me was seeing those fresh pork hams hanging from the rafters of their huts just covered with flies.

When I took visitors through the village, I would pull back the locks of hair of the kids. There were always millions of lice and nits around the edges of the hair. Nits are the eggs of lice, and they were laid around the roots of hairs. But lice didn't seem to bother those kids.

When we made our tours, the Indians liked to show off their handiwork. A lot of their crafts were made from the local trees and shrubs. Visitors really seemed to like the tours, and everybody learned a lot from them, or at least I did.

Their huts were built by driving six posts into the ground and then attaching a solid wood platform about two feet above the ground. The Indians slept on the hard floors. The roof was made out of palm fronds that overlapped.

It didn't take me long to figure out which of the Indian families were the richest. The number of beads hung around a squaw's neck told the story. Many beads was a sign of a lot of dough.

The squaws hardly ever spoke to strangers. They just

sat on the ground with their legs crossed, holding little sewing machines in their laps, or they sat on the edge of the wooden floors of their thatched palmetto huts. They spent all day sewing jackets, pants, dresses and shawls. Although they never spoke to strange men, they would keep their eyes glued on their chief, Billy Bow Legs, whenever he was around.

 The Seminole men were always friendly. Chief Billy was always especially friendly to me. He liked to see me pole canoes. I was the only white man he knew around Sparklin' Waters who could pole a dugout. It is kind of tricky. You have to use a long pole to push the canoe along, and you have to stand in the right position on the little platform near the back of the canoe or the boat will reel. If it rolls, it spins just like a rubber ball. Still, those Seminoles would go gliding down the river in one of those boats with a squaw and three or four kids on board.

 A few months after I started leading tours through the reservation, the chief went back to the Everglades, which was his home. About a year later, he sent me an invitation to come visit him. He had hewn a dugout out of a cypress log and was going to give it to me, but I never got down there. I always regretted that. I liked to show women around Sparklin' Waters using my old canoe, but I know a new hand-hewn dugout would have looked a lot more genuine. It would have been the real thing.

15

"**HOW WOULD YOU** like to go out hunting gators tonight? We'll glide down the river," Rob said.

It was less than six months after I came to Sparklin' Waters that Rob Allen invited me to go alligator hunting with him. He liked to hunt all kinds of alligators and snakes, but newborn gators really turned him on. He liked me because I was always ready for some new excitement.

Rob lived in a house he had built out of concrete blocks all by himself. The house was big enough to have separate rooms for his two daughters, and there was a big bedroom for himself and the woman he was about to marry. Although the house had most of the conveniences to be had in those days including an indoor toilet and bathtub, his new wife still had to build a fire around a kettle hanging in the backyard to wash clothes. She scrubbed their clothes on an old washboard. Then she hung them out to dry on a clothesline.

Rob kept a bunch of caged animals in every room in his house. But what really made him stand out was a snake pit that he built in his front yard. He managed to keep more moccasins there than I had ever seen. That was the hub of the Florida Reptile Institute, which he developed soon after he got to Sparklin' Waters.

When Rob asked me about going hunting for gators with him, I jumped at the chance. "I'll be ready any time after five-thirty," I said.

I got to the river's edge just as the sun was setting. I was ready to get started. Within minutes, Rob drove up in his old station wagon. It was loaded. His sister, Eleanor, and Margy, the girl he was about to marry, were in the car with him.

Rob's sister was a beautiful girl, and I liked her, but I was not expecting to have any women go on that hunting trip. "You're not gonna take these women with us, are you?" I asked.

"Sure, we're all going." he answered.

"Gawd Aw'mighty! How are the four of us going to fit in a sixteen-foot canoe?"

"That's easy."

Rob took charge. "Newt, you sit on the back seat."

I was just finding my place to sit when he added, "Now take this paddle—and spread your legs.

"Eleanor, you sit on the floor in front of Newt, between his legs—that's right. Now, Margy, I want you to sit on the front seat—that's right."

"Now, I'm gonna kneel down in front so I can watch where we are going. Newt, I want you to paddle downstream. But don't let the paddle touch either side of the canoe. Alligators can sense even the slightest vibration over a mile away. I'll signal which way I want you to go," he said.

Rob pushed off the canoe, hopped in and knelt down in front of Margy. We were off and running. Rob was looking ahead with a spotlight shining from his cap so he could see anything in front of us, and by turning his head side-to-side, he could keep an eye on both banks of the river.

I paddled the canoe just like Rob said. He signaled

with his arms. When he pointed to one side, I paddled the boat that way. When he raised his arms above his head, I slowed the canoe, and when he put his hands to his sides, that was a sign to keep moving straight ahead.

    We must have gone about half a mile downstream when Rob shined his light across the river and saw at least twenty-five pairs of fire-engine red eyes glaring back at him. He pointed them out to us. Slowly and quietly, I steered the canoe toward those red eyes. Each pair belonged to a baby alligator. The mama alligator watched with her eyes staring at us from above her brood. I eased over to the bank. As Rob reached to grab her, she slithered into the water. When she dived in, Rob plunged in after her and brought her back up. She was at least five feet long, and she whipped her tail from one side to the other, but Rob held on. He rolled her into the canoe and tied her feet across her back. While holding her head, he wrapped a rope around her snout so she couldn't get her mouth open. Then he forced her into a sack. He laid her on the floor of the canoe and collected all the baby alligators in another sack. That was a good catch for our first night out. In fact, that was a good catch for any night. We paddled back home gloating over our success.

    After that, Rob and I hunted often in the Okena National Forest. It was teeming with all sorts of bugs and small wild animals. The hunting trips always began with a call from Rob saying, "We can go hunting tonight. I'll pick you up at seven o'clock."

    I always agreed unless there was something else going on that I couldn't get out of.

    One night when Rob drove up to the riverbank, he had three fishing poles with five-foot lines.

    "What are we gonna do with those things?" I asked.

"We're gonna catch some skinks," he replied.

Skinks are lizards with long tails that thin down to a point. Grown skinks usually are about twelve to eighteen inches long, overall. They live on insects.

We slogged through the jungle, past bushes and through briar patches as high as our heads. Rob kept shining his light up into the tree branches. He suddenly spotted a skink, a foot and a half long. Then he tied a piece of bacon rind on the end of a line hanging from a fishing pole and dangled the bacon above the skink's head for only a few seconds before the skink took hold. With the bacon rind in his mouth, the reptile grabbed the line. Rob snatched the animal's feet free from the tree and dropped him in the sack. Rob never gave skinks a chance to bite him although they are harmless.

16

ROB'S REPTILE FARM was one place that tourists were told about. A lot of them liked to go look in the snake pit, I think mostly out of curiosity. Very few stayed for long except on Sunday afternoons when we were milking rattlers.

Rob's interest in snakes, especially poisonous ones, really stood out. I learned about his interest when he called me one Sunday morning, "Newt, would you like to come over and help me milk some rattlers this afternoon?" he asked. That was almost a year after I started working at Sparklin' Waters.

I had seen the snake pit Rob built in his front yard but had never gotten up close enough to see inside. It was twenty feet square, fenced in by a concrete wall three feet high. There was a ledge outside the wall for kids to stand on so they could watch the snakes being milked. The inside of the wall was painted with slick enamel so the snakes couldn't crawl up it. Around the inside of the wall was a moat fourteen inches wide and six or eight inches deep. It was filled about halfway with water.

Inside the pit, Rob had two hundred rattlers that were coiling, hissing and rattling all the time. Some of the snakes were fully grown, but there were also a lot of young ones. The very sight of the pit was enough to scare the living daylights out of me the first time I looked down in it. I never even dreamed of picking up one of the scary monsters. Ever since I had lived on the farm in Georgia, I had been scared to death of poisonous snakes, especially rattlers and moccasins. I had never handled them while they were alive without a stick or a pole.

    My first experiences with Rob showed me that he knew what he was doing in the wild. We had collected all those insects, skinks and even large ferocious alligators. I realized that Rob knew how to handle the snakes, but I also knew that I didn't have any idea how to milk those things and I had to get training. I needed lots of teaching.

    Every Sunday afternoon, Rob jumped into the pit and milked the snakes. Rob warned me starting my first time to wear bite-proof leather boots that came up to my knees. When I got to his house, Rob jumped in the pit and started to show me how to pick up the rattlers and milk them. He explained how to grab the snake behind the head and hold it to hang its fangs over a vial. Pressure on the back of the snake's head caused it to squirt the yellow venom into the vial. The venom looked just like orange juice.

    Once I learned how to handle the rattlers, I jumped in with Rob every Sunday afternoon whenever I was free. We got several flasks ready ahead of time. When we climbed into the pit, we caught the snakes and held them over the rim of the vial. Occasionally, a snake's fangs got knocked out, but new ones grew and took the place of the lost teeth. We usually filled a couple of teacup-sized vials in an afternoon.

    We dried out the venom in a double boiler, causing it to crystallize. Then the flakes were scraped out of the big flask and dropped in small vials. Each vial held one ounce. Rob charged twenty-five dollars per vial. Crystals of venom were injected into horses to get antivenom.

Working with Rob, I learned a lot about snakes. I learned how to show them off. I learned how to make money selling the venom, but most important of all, I learned which were poisonous and which were not.

And I was lucky. I was never bitten while I was working in the snake pit.

17

"**WHAT ARE YOU** doing holding that alligator down out here on the beach?" asked the visitor.

"We're about to make a movie, wrestling this fellow. A lot of it will be underwater, but you can see everything from right here on shore. You can stay and watch if you want to," I said.

I was trying to explain what was going on to a stranger who just happened to walk by the cove at the time. I was holding Cannibal's snout with both hands. Cannibal was a sixteen-foot alligator. Rob Allen was going to wrestle him. Rob was standing on the pier, and two cameramen were getting set up to shoot the whole affair. Rob was waiting to dive in as soon as I let go of the beast.

AN ALLIGATOR WRESTLER has to dive in and climb on top of the gator. He grabs its left front leg with his left hand and keeps the snout closed with his right hand. Then he throws the gator across his chest. When the alligator rolls, the wrestler rolls with him. The wrestler has to grab a breath every time he gets his head above water. Rob had lots of experience. He wrestled alligators over and over. This time, the stunt came off just like we planned it.

When all the pictures were finished, the stranger

blurted out, "That wasn't much." Then he left.

Later, he came back and asked Rob, "Would you let my stuntman wrestle that alligator there? He's got a lot of experience diving and fighting octopuses and sharks and a lot of other animals out in the ocean."

Rob always tried to satisfy everyone. He explained that the only requirement was to note the following on any film produced: "These photographs were made at the Florida Reptile Institute, Sparklin' Waters, Fla." He added, "Besides scheduling the event, that's all you have to do."

They agreed. The stranger was going to bring his stuntman back to wrestle the alligator. When the stuntman got there, I held the gator.

The stuntman looked me and the alligator over. At the time, I was tall and skinny. I weighed only about one hundred and fifty pounds. "Do you think I can wrestle that big animal?" he asked.

"Sure, fellow, there's nothing to it. I'm sittin' here holding him with just one hand," I said.

"Yeah, but you know what you're doing." he replied.

The cameramen, Rob and I all got ready. The stuntman stood on the platform of the pier, and I let the gator go. He slithered into the water. I yelled, "Man, you'd better get on that gator's back. Get off the dock! Get after him!"

"I can't do it. I can't!" he yelled back.

I turned to Rob and shouted, "You better get that gator before he gets in deep water, or during the night we'll have to hunt for him downriver!"

Rob dived in, caught the alligator and, as he brought him out, commented, "There's nothing to it."

The stuntman asked, "How much do you charge to

wrestle him?"

"Five hundred bucks."

Rob had already wrestled the gator once. So I wrestled him one more time so the man could see how simple it was.

It was six months before the stuntman came back to the Florida Reptile Institute. By that time he had built up enough courage to have movies made while he was wrestling a monster.

The lot across from the creek was hemmed in by a canal and a chain link fence. Horses and cows grazed while hogs wallowed in the mud along the edge of the canal. Alligators lolled on the grass, just waiting to seize some fresh pork. Suddenly, a pig squealed. He was in the grip of an alligator. The formerly frightened stuntman jumped at the chance to have a picture made of himself rescuing that poor pig. He told the cameramen to focus on him while he was saving the life of that squealing animal. With the cameras focused on him, the stuntman jumped the fence and landed on the bank of the canal. Unfortunately, he was wearing old tennis shoes, and there was no sand to give traction in the muddy field. When his feet hit the mud, they slipped out from under him and he flipped. The alligator was waiting. With one snap of the jaws across the stuntman's belly, the alligator ended that rescue attempt.

I could only reckon, "I always tell everybody, don't ever steal another man's trick unless you know what you are doing because you can surely get in trouble!"

IN SPITE OF a lot of difference in our backgrounds and basic interests, Rob and I admired each other, and we developed a close friendship. We frequently hunted together and collected insects and many other animal specimens, which he advertised and shipped all over the world. My admiration for Rob and his attachment to me lasted through the years long after I left Sparklin' Waters.

Many years later, when I opened Chin-ka-pin Ranch, Rob moved to Floridaga and opened another reptile farm. He came out to my place every chance he got. In fact, he lived in town until he died at the age of 73. He was a great man.

18

"BROOKS! BROOKS, STOP! Can you get me a forked stick so I can pin that rattler down?" I asked.

Brooks jammed on the brakes and climbed off his perch on the gas tank of the old flatbed farm truck. He found a pine limb that divided into two branches lying on the side of the road. He trimmed it into a fork and handed it to me. "Here, try this," he said.

It was after I had decided to leave Sparklin' Waters and while I was working my final week there that I heard Daddy had died all of a sudden, apparently from a heart attack. There was nobody but my brother, Brooks, and the hired men to do the work on the farm, so I bought a bus ticket to Ashburn and went to see if I could help Mama with the farm. There was too much work on the two farms for Mama and Brooks to handle by themselves.

That day when I yelled at Brooks to stop the truck and get me a stick, he and I were hauling a load of lumber from a sawmill near Isabella over to the farm outside of Ashburn. Brooks was driving and I was sitting on top of the lumber. The cab of the truck was cut away, so the driver just sat on a gas tank. When we rounded a curve about a mile east of Coverdale, we met a lady and her kid walking along the side of the road. Just about ten feet in front of her, a big rattlesnake was wriggling out of a path running from a cornfield onto the paved road. Sitting up as high as I was, I saw the snake first and yelled, "Stop!" The snake coiled. He was ready to strike.

I jumped down from the pile of lumber just before the snake could strike and kicked sand on him. The rattler slithered into the grass on the other side of the path and coiled again.

As soon as Brooks handed me the stick, I pinned the snake's head to the ground and reached down and grabbed him behind the head just like I had done s of times in the snake pit with Rob Allen. Then, I wrapped the body and tail around my arm and started to get back on the truck. Just then, a neighbor came by and yelled, "Put that thing down! You'll get bit. Don't hold him that way!"

"Oh that's all right. I've done this lots of times. When you hold'em right, there's no way they can bite you," I explained.

With those words, I climbed back up on the lumber and held the snake with my hand until Brooks and I got home. The snake rattled all the way. I put my new trophy in a pen.

A couple of weeks later, I was looking after a work crew on the farm west of town. I had gone home to get lunch for the men. On the way back to the farm, when I was about a mile and a half from Cousin Bill Whiddon's tomb, I saw another rattlesnake, between five and six feet long, crawling across the road, just twisting and turning. I jammed on the brakes and jumped out of the truck. I grabbed the forked stick I had left in the truck after my first scuffle and trapped the snake's head. I grabbed the snake behind its head and climbed back onto the truck. I held that snake hanging over the side of the truck while I drove the mile and a half to Cousin Bill's house. The snake just kept on wriggling and rattling all the way. When I honked the horn, Cousin Bill's daughter came running out on the porch. She heard the snake rattling, even though she couldn't see him.

"What you got there?" she screamed. "Put that thing down!"

"Don't get excited. I just got a big diamondback rattler. Could you get me a croaker sack, one without any holes in it?"

When she came back with the sack, I dropped the rattler in the bag, tied it and threw the bag in the back of the truck. I then drove on to the camp house. When I got there, I found a prison crew working on the road out in front. Two officers were standing in the road guarding six convicts in shackles. When I stopped, one of the convicts yelled out, "Boss, Boss, I hear a rattlesnake!"

"Don't worry, fellows. I have him in a croaker sack," I said.

I picked up the sack, untied the top and dumped the snake on the ground. The snake coiled and did he ever rattle! Then I pinned the rattler with my forked stick and put him back in the bag. I carried him home and caged it with his cousin.

Two days later, while I was transferring a pregnant female from one pen to another out in the back yard, I was holding the rattler with my left hand. The snake gave a quick turn and bit my left thumb.

"Did that snake bite you?" Mama asked.

I didn't answer. I just dropped the snake between my feet and made her coil. I asked Brooks to run in the house and get a new razor blade. With that blade, I slashed each fang mark in my thumb forming deep "X's." My finger bled a lot. I let the blood drop on the coiled rattler's head.

I never even went to the doctor. My arm ached for a couple of days, but I never felt another thing.

I shifted the snake into another pen. About a month later I took both snakes to Sparklin' Waters and sold them to Rob Allen's Reptile Institute for a dollar and a half each.

19

"NEWT. NEWT. NEWT!" screamed Eileen over the megaphone she ordinarily used to call Rob Allen when he was at home. That was half a mile across the woods from the ticket office.

"Newt, Newt, Newt! Hurry, Hurry! I got some swimmers!"

Just by the sound of her voice, I knew Eileen had something big. What could it be?

Eileen was a receptionist at Sparklin' Waters. Her job was to sell tickets for rides on the glass-bottom boats. But this was a gloomy Sunday afternoon, and nobody was swimming and nobody was buying tickets to ride on the glass-bottom boats. I was doing lifeguard duty, and I had about decided that there was no use sticking around, so I had gone into the bathhouse to take off my sticky bathing suit.

When I heard Eileen yelling, I pulled up my bathing suit and ran back to the beach. When I got out in the open, I saw her running up the beach. She was still yelling into the megaphone. When she saw me and knew she had my attention, she stopped running and quit yelling in that blasted amplifier. And when she knew I would help the visitors, she went back to the ticket office.

A man and two women were strolling across the sand beach. They had driven up in a new robin's egg blue convertible La Salle. Slim Garner, dressed in a full-body bathing suit, was stepping briskly, walking ahead of the two women. He was a tall, thin man, probably about six feet. As he came near the bench in the middle of the sand beach, he yelled over to me, "Howdy! We want to go swimming. Is it safe?"

"Go ahead. There's no lightning," I said.

Although Slim Garner's words could be easily understood, they were accented. It was a soft accent which I couldn't place at the time. I just knew it was not Southern, nor was it from anywhere in the United States that I knew. I also knew the man was trying to act like an American. He must have learned the way he was acting from some of the "Wild West" movies.

One of the women was trailing just a few feet behind him. She had taken off her shoes and was digging her toes in the sand. She was Ethyl Crutchfield, a cute little lady with dark brown hair. She was wearing a colorful kimono, tied at the waist by a wide sash. When she got near the bench in the middle of the sand beach, she untied the sash and threw the kimono and sash across the back of the bench. Then she and Slim made a dash for the water. They splashed in and kept running until the water was so deep they couldn't run any more.

 The other woman, Marie, was straggling along behind the others. She was more robust and had blonde hair that was obviously tinted. Both the women spoke perfect English, at least American English.

Marie was wearing a cotton print dress with a pair of white sandals, which seemed to blend in with the sand on the beach. She was holding a two-quart Thermos jug in one hand and four waxed paper cups in the other. By the time I got to the bench where Ethyl had thrown her kimono, I whirled around and saw Slim and Ethyl in water up to their waists. They were jumping up and down, but they didn't appear to be in any danger, so I turned my attention to Marie, who was sitting on the bench by then. The bench was at the back edge of the sand beach, near the grass lawn. I sauntered around behind the bench, so I was facing the water and could keep an eye on the swimmers while I was talking to Marie. I put one foot over the back of the bench and leaned forward so I could see her face when I spoke. "How're you, lady?" I asked.

"Fine," was all that came out. She swung the Thermos around and poured a paper cup full of whatever she had in the jug. Then she snapped, almost in a sneer, "Who are you and what do you do around here?"

"I'm the lifeguard! I am here to save your life if you need it!"

Marie suddenly became serious, but she sounded rather flippant when she blurted out, "Oh, you are just the man we need to know! How about a drink of pink lemonade?"

"Naw, thank you. I don't drink no kind of whiskey—nothing with alcohol."

"You're just the man we need. We need you bad," Now, she was getting serious, really serious. She was almost begging.

I later learned that Marie was drinking Dubochet sloe gin mixed with lemon juice, but it was mostly gin. She was guzzling it just like it was drinking water.

I kept watching Slim and Ethyl, who were cavorting about halfway between the beach and the outer limits of the swimming area. After a few minutes, they dashed out of the water and across the sandy beach to the bench where we were. They each poured themselves half a cup of "lemonade" and gulped it down. They ran back across the beach and splashed into the water again but came back again and again. Every time they came up to the bench, they gulped down another half cup of Marie's pink lemonade.

After nearly an hour and a half, the three of them gathered around the bench and gulped down the rest of the lemonade before Slim and Ethyl grabbed their towels and dried off. Ethyl slipped on her kimono and tied the sash. They were obviously getting ready to leave. All of a sudden, they turned and asked me, "Where do you live?"

"Down the river, but I change clothes here in the bathhouse."

Marie and Ethyl, almost in unison, said, "Come on and go with us." Slim chimed in, as well.

I answered, "I can't leave here like this."

"Just go in the bathhouse and put on some clothes and come with us," they said.

Since by then it was almost six o'clock and there were still no swimmers in sight, I was free to leave, so I trotted over to the bathhouse and put on my khaki pants and a slipover shirt. When I got back to the bench, they had collected their belongings and had started walking toward the parking lot.

I tagged along behind them as they moseyed over to that blue La Salle. When we got to the car, Slim ordered, "You drive since you haven't had anything to drink."

I hardly had time to say anything before Marie and Ethyl, almost as a chorus added, "Come eat supper with us. We need you to help us find a place."

"Okay. I'll find some place," I said.

I opened the doors on the driver's side. Marie and

Ethyl crawled in the back. Slim went around to the passenger side and climbed in the front seat. I slid in behind the steering wheel. I had to look around to find my bearings. Slim pointed out where the switch was, then the starter. I was familiar with the clutch, gearshift, brakes and gas pedal. I was in hog heaven, about to drive Marie Critchfield's brand new La Salle convertible.

    I drove over to the square in Okena, where Riley's, a little Mom and Pop restaurant, was open. It was near the crossroads in the center of town. We sat around a little table in the middle of the restaurant. There was only one other couple eating at the time.

    After we sat down, a middle-aged lady with her graying hair tied in a bun behind her head came over to the table with four glasses and a pitcher of water and four menus. When she set the glasses down and filled them with water, she commented that she could take our orders as soon as we had time to look over the menus. It was hardly more than a minute before she returned to ask in her soft, perhaps slightly hoarse voice, "May I take your order?"

    Marie and Ethyl asked for salads, but we were quickly told that about all they had on Sunday evenings was soup and sandwiches. If one wanted a hot meal, they had fried chicken served with rice and gravy. A side dish of baked beans was also available.

    We all had a bowl of hot vegetable soup before the fried chicken. We had the choice of dark or light meat. Both Marie and Ethyl had breasts, and Slim and I got thighs and legs. Wings were thrown in for free. After the chicken and beans, we had some slices of chocolate cake, and each of them had a cup of coffee. As it turned out, the meal really tasted pretty good, at least to me, but they were not happy with the cake. It just didn't taste "like French pastry."

While we were eating, I learned that Marie and Ethyl were sisters living in Lexington, Ky. They had the Coca-Cola franchises in twenty-nine counties in the area.

Marie and Ethyl had visited Paris in 1920. That was when they met Slim. He had several sightseeing tours in Paris that he started after World War I. By the time they met, he owned five fleets of limousines. He made several trips to the United States after he came to know Marie and Ethyl. They usually got together in Jacksonville and visited with a mutual friend, Charlotte Buford, whose husband, Harry, was a stock broker. This time, they had decided to come directly to Sparklin' Waters just to see what it was like. They had rooms at the Sparklin' Waters Court. On later visits, Charlotte and Harry came with them.

Slim was 42 years old and stood tall. He was a little muscular, but he certainly did not threaten Atlas' supremacy. He had a full head of dark brown hair which was graying around the temples. He was always full of life like he was that afternoon in swimming with Ethyl at Sparklin' Waters. He liked to drink, but I never saw him get stewed. I never noticed him stagger, and he was always able to carry on an intelligent conversation. His French accent—after I recognized what it was—always was present. He was taking off the whole month of August for his vacation.

After we ate, I chauffeured the three of them back to the hotel. They invited me in, so I listened to them gab for about thirty minutes before I blurted out, "I've just got to go home." It was already nine-thirty and I had to swim back to Camp Nasty. Slim drove me to the bathhouse where I redressed in my lifeguard's bathing suit, my uniform.

I then plunged into the river and swam to the float near the guard tower where I peeled off my bathing suit. I swam down the river to the jetty in front of Camp Nasty in the altogether. The night was pitch black, and it was almost impossible to see more than three feet in front of me. I swam down the middle of the river. The outlines of the treetops against the sky made it possible for me to know where I was. I was able to make out the jetty, I really don't know how. When I climbed on the dock, the headlights of a parked car flashed on me. I walked directly toward the headlights, still in my birthday suit. When I got to the car, I recognized the couple sitting in the front seat. They giggled, but I ignored their laughter. I stopped and joked with them for a few minutes. They asked why I was dressed so warmly. I just said that I liked it that way and that there was less drag when I swam down the river without clothes than when I wore my bathing suit. After that, neither they nor I mentioned my dress or lack of it. After a bit of gab, I continued to climb the hill to Camp Nasty.

The next morning, I splurged for breakfast. I had two slices of bread and a glass of orange juice. After that I dived into the water and swam upstream. When I got near Sparklin' Waters, I saw a few people on the beach. There was a crowd over by the ticket counter. Swimming up to the float, I reached underneath to get my bathing suit. After I was able to squeeze into it, I swam to the sand beach and walked over to the bathhouse, dried off, and put on my brogans. I was back in uniform and ready for a new day's work.

It was about one-thirty in the afternoon when I looked

up and saw Marie's La Salle come rolling through the driveway and into the parking lot. Slim was driving. All three visitors were sitting in the front seat. They got out and sauntered over to the sand beach. They walked straight to the bench behind the center of the beach and took it over for themselves, a place to sit and shed their outer clothes.

Before they sat down, they looked around to find me. I was in the lifeguard's tower. They waved. Slim was dressed in a bright blue shirt with an open collar and white pants, and Ethyl had on a brightly colored print cover-up, which she shed in about the same time it took Slim to slip off his shirt and pants. Once again, Marie stuck to the bench, at least for a time. All three were drinking bottles of Coke as they walked across the beach, but Marie put the empty bottles in the satchel she was carrying. I knew without being told that she was carrying another Thermos of pink lemonade in her satchel.

Slim and Ethyl ran back to the swimming area. Marie, who was wearing a tan cover-up, doffed it to expose a brightly colored one-piece bathing suit. Then, she strolled over to the water's edge to test the water with her big toe before wading in. Then she dived in and swam over to the tower where I was perched. There she called out, "How're things?"

"Fine. Did you have a good night?" I asked.

"Sure did. Hope you'll show us some more of Okena tonight."

"Sure 'nough, but I can't leave here until after five thirty."

"That's all right. Are you going to stay up there all afternoon?"

"Oh, I can come over to the beach from time to time,

but I just can't leave the swimming area."

Again, I looked up to see Slim and Ethyl jumping up and down in the water. Then Slim swam out to the edge of the swimming area and back toward the beach. Ethyl was right behind him. They were both good swimmers.

Marie swam back to the water's edge and met them. They walked over to the bench, and each one took a big swig of lemonade. Then it was back to the water for Slim and Ethyl.

I climbed down from the tower about two-thirty and ambled over to the bench where Marie was sitting. We ventured into some small talk before Slim and Ethyl joined us. There was still more small talk before Slim spoke up. "Newt, can you find us a nice place to eat tonight?"

"Sure. I'll try," I said.

With that, I began running the different cafes in the area through my mind.

"You might like the Raywood Drive-in. It's a couple of miles south of town and run by Mrs. Renee Woods."

Renee was known around town as Madame Dumas, but I didn't mention that. She was a native Parisian. All three of the guests accepted my suggestion without asking for more details.

"Can you take us there?" they asked.

"Sure. What time would you like to go?"

"After you get off—probably around six or six-thirty."

"Fine. I'll be waiting," I said.

After that conversation, the three of them drove off, and I didn't see them until just after six o'clock when the convertible came rolling up the driveway. Slim hopped out, and he and Ethyl climbed in the backseat.

Marie stayed in the front passenger seat. That left only the driver's seat for me. I was back in hog heaven. I was about to drive that convertible limousine again.

I drove around the town square before heading south on Highway 441. It was still early, so I was able to park near the front door. When Madame Woods opened the door to welcome us, she hardly had to say more than "We're glad to have you" before Slim recognized the accent.

"Pardon, Madame, mais n'etes vous pas francaise?"

"Bien sur, Monsieur."

After the greetings, Madame Woods almost ran into the kitchen to prepare a French meal. There was wine, lots of it, both white and red. There were hors d'oeuvres and a salad followed by roast beef, mashed potatoes and string beans. We also had some brie, a soft cheese, and coffee which looked like that served in most restaurants, but I never tried it. I was sober, but my guests were a little tipsy. Slim and Renee spoke French all evening. I didn't understand a word of it, but the language really rolled off their tongues, and the sound of it was beautiful.

Just before we left, Slim spoke up, "Serait-il possible de reserver tout le restaurant pour demain soir?" he asked.

"Oui. Oui. Bien sur, Monsieur."

Slim translated their conversation for me, so the next night was already planned.

When we arrived for our second dinner there, Slim stood by the door and turned away three or four guests. As soon as we got inside, Madame Woods served an aperitif, whatever that is. It's alcoholic, so I turned it down. She also had a plate of hors d' oeuvres, more finger foods which included stuffed mushrooms, a few vegetables and, oh yes, some shrimp, but they called them prawns. For me, they were still shrimp. They were really good. She gave me a glass of orange juice. When we sat down at the table, there was more wine, both red and white. We were treated to another elegant dinner! There was fish and pheasant, served over rice with side dishes of carrots and yellow squash. The bread was sliced baguettes, fresh and hot with a soft center but a hard crust. After that, there was a green salad, followed by fresh slices of apple intermingled with some beautiful grapes and some orange slices. Would you believe, there were eclairs served with the strongest coffee I ever tasted. I swallowed only a sip, and it nearly blew my head off. The meal was fixed in a way that no country boy had ever seen!

    After those first two nights, I became a sort of caretaker for Marie, Slim and Ethyl whenever they visited Sparklin' Waters. When Marie drank too much, she handed me her diamonds. I also became her banker when she was carrying a lot of money. I really didn't like to hold all that loot. After chauffeuring them to the Sparklin' Waters Court, I went back to the bathhouse where I hid the money and jewelry in my clothes locker. The day after a night on the town, the three of them came by to get back their valuables.

    It was on the first afternoon of their next visit that Slim handed me fifty dollars and asked me go to the cab stand in Okena and buy booze for the night. He wanted imported liquor. Those were prohibition days, and any liquor was illegal. Strong alcoholic beverages, even though bonded in another country, were considered bootleg. Marie's favorite was Dubochet sloe gin. She always called it "pink lemonade."

    Every time Slim, Marie and Ethyl visited Sparklin'

Waters, they reserved the same three-bedroom apartment. They always had something to celebrate.

The group kept inviting me to visit them over in Jacksonville. Finally, I agreed. Marie reserved a large suite on the ground floor in the old Windsor Hotel facing Hinning Park in downtown Jacksonville. A porch stretched across the front of the old wooden lodge.

Slim stayed at the George Washington Hotel a couple of blocks away. Harry Howell's Restaurant with its back door whiskey shop was located diagonally across the street to the east. The restaurant was one of the best. The waitresses were fantastic!

Slim invited me and Diz Thomas over to go to a Georgia-Florida football game, which was always played in Jacksonville on the last Saturday afternoon in October. Diz liked his "hooch," so before leaving Okena on Friday afternoon, he went by Milton Saunders' bootleg shop and bought two gallons of Milton's best red moonshine.

We drove to Jacksonville in the little panel-bodied Austin-Healey that had dioramas advertising Sparklin' Waters on each side of the body. The dioramas showed fish swimming through the greenery at the bottom of the river. It was a dazzling picture.

We got to Jacksonville about five o'clock Friday afternoon and parked the little truck with its exhibits between the curb and sidewalk on the corner of Laural and Adams streets. After locking the truck, we checked into the George Washington Hotel. The police tried to find Diz to arrest him for parking the truck on the street corner, but they never found him. We stayed out of sight and left the truck right there where it attracted the view of lots of spectators. It was doing its job, advertising Sparklin' Waters.

The lower floors of the George Washington Hotel were solidly booked when we checked in, so we were given a room on the top floor where the Georgia football team was staying. The team members made so much noise that we couldn't sleep, so we joined the celebration. It lasted all night long. At six-thirty Saturday morning, Slim and Diz and I were still celebrating and still gulping down booze. That was the last I remembered until about six o'clock Saturday afternoon when I woke up to find myself in bed with one of the most beautiful women I have ever seen. We were both stark naked!

My first thought was the ball game, and I asked, "When does the football game start?"

"Oh. The game was over two hours ago," she said.

My thoughts then turned to the money in my pants, but I was distracted by the need to pee. I hopped out of bed and ran to the bathroom. I glanced through the door into the next room where the bed was empty.

When I got back, the doll in my bed was wide awake. I asked her, "How in the world did I ever end up with you?"

"Sally and I are night waitresses at Harry Howell's Cafe. Sally is with Slim. When he drinks too much, we take him home to sleep it off," she explained.

I was really lucky. My money was still in my pants pocket. I guarantee you, that was the only time in my life I ever made an ass of myself by drinking too much. I made up my mind right then and there that I would never take another drink of alcohol. We went out that Saturday night and had a great time, but I never drank another drop of liquor!

Slim went back to Paris, but Marie moved to Sparklin' Waters.

Several years later, when I was driving an interstate bus, Marie moved to Miami. She asked me to dinner. I was impressed with her bay-front, two-story home on Brickle Avenue. The view of the bay from the house was spectacular. Still a country boy, I was even more impressed at the dinner table when Marie rang a tiny bell for the maid to serve another course. I still remember that simple gesture. I visited several times but I hated her two Great Danes that roamed all over the place. Marie asked me to marry her several times, but I balked. I later thought maybe I used poor judgment. She eventually married a local man about my age. I learned she died of cancer.

20

"NEWT, CAN YOU make home brew?" asked Diz.

"I used to help Mama make it for Daddy when I lived at home. But that's a long time ago," I said.

Diz Thomas was feeling me out one afternoon after we got off from work. It was just about five-thirty, and we were walking along a garden path toward the river.

Diz is a short, potbellied man, about ten years older than I am. He was hired to help do odd jobs around Sparklin' Waters and to drive the Austin-Healey truck with its graphic pictures of the park mounted on each side of the panel body. He drove the truck through the towns in North Florida as a means of advertising the resort.

Diz still liked his hooch, but I swore off the stuff forever after I passed out the day of the Georgia-Florida football game in Jacksonville. We were both bachelors and got to be good friends. At least I thought we were good friends until he refused to repay me for taking on his chores while he took a vacation. Eventually, that led to my decision to leave Sparklin' Waters.

After he asked me whether I could make home brew, Diz got to the point. "If I furnish everything, would you try to make some for me?"

"I'll give it my best shot. Sure will," I said.

"What do you need?" he asked.

"I'll have to have a case of thirty-two-ounce cans of

malt and a case of five-pound bags of sugar, some yeast and some beer bottles. I'll also need a couple of five-gallon jugs, a rubber hose about four feet long to use as a siphon, and something to cap the bottles with," I explained.

The next night when I got to Camp Nasty, I found three barrels of beer bottles in the back yard. A case of malt and a case of ten-pound sacks of regular granulated sugar were inside the shack, sitting next to the old stove. How he got that case of sugar in without some help, I'll never know. There was a box of yeast sitting on the stove along with two five-gallon jugs, two rubber hoses, and a gadget to cap the bottles. I knew right then that I had a job. I was in business—the moonshine business!"

After mixing a can of mash with five pounds of sugar in each of the jugs and throwing in a pack of yeast, I added enough water to fill the jugs, topped them with balloons and set them in the back room. I took a look every day. After a couple of weeks, the balloons blew off, and I could see bubbles rising to the top of the jugs now and then. I siphoned the top liquid out of each jug into the regular beer bottles. I capped each one of the bottles and took up the boards that covered the hole in the floor of the front room. I set the bottles of fresh brew underneath the house in the dark and left them for over a month. After the brew aged, I called Diz over to the pier where I was working on the glass-bottom boats. "I got some home brew ready, just waitin' for you!" I said.

Diz couldn't wait. "I'll tell you what I'll do. I'll bring my girl friend and I'll bring you a girl. We'll be over at Camp Nasty tonight about seven-thirty," he said.

"Could you bring over a tub and some ice to chill the brew sometime this morning?" I asked.

At lunch, I swam home and poured the crushed ice into the No. 2 washtub Diz had left. I then screwed fifteen bottles of brew down into the ice and sprinkled some salt over the ice. I added another bag of ice after I got home just to be sure the drinks would be ice cold. When Diz and the girls arrived, the drinks were just right.

When they got to my shack, Diz and the girls had to sit on some wooden apple crates I dragged in while they gulped down the beer. Diz got drunk as a coot, but I stayed on the wagon.

After that Diz made a regular showing with a lot of other folks, just to party. That went on for months.

"MR. ED, WOULD you please bring me three bags of nice grapefruit and drop them on the bench when you come over here this evening?" I asked Ed Masters, who, besides owning the land where Sparklin' Waters was, owned the Sparklin' Waters Court. He also owned a large grapefruit farm downriver. It was getting near wintertime and the grapefruit were just beginning to sweeten.

"You know it, boy, You know it. I'll bring it!" he said.

"You know it, boy! You know it" was Mr. Ed's favorite expression when he was talking.

When I finished my work that day, I collected the six bags of grapefruit he had left—beautiful grapefruit! He brought twice as much as I had asked for.

"My goodness alive!" I didn't say anything out loud, but that was my thought.

I loaded my canoe with three bags and paddled downriver to my dock. After tying up my canoe, I lifted the bags onto the wharf and carried them, one bag at a time, into the cabin. I had to go back to Sparklin' Waters for the other three bags. When I got them all into the cabin, I sliced the grapefruit across the middle, took a half in each hand and squeezed the juice into a No. 2 wash tub. I strained the grapefruit juice, getting out all the seeds and pulp. There was enough clear fruit juice to nearly fill both five-gallon jugs. I dumped in a few pounds of sugar and a packet of yeast. To this day I can't remember how much sugar I put in there! I put a rag on top of each jug and left the jugs to ferment. I looked at the mash every day until I could see bubbles rising up through the juice, not constantly, but one bubble at a time. When I saw that, I said to myself, "She's ready to come off!"

I siphoned the liquor down to the sediment on the bottom. That was about an inch deep. The liquor was divided among a dozen beer bottles. I capped the bottles and set them under the house. I didn't tell Diz about it until after they had been there for about three months. Then I invited him: "Diz, come out tomorrow and bring some special girl friends because I've got something special for you."

"Gee, I thought we had drunk up just about all the brew you made. But yeah. We'll be there," Diz said.

In the meantime, I selected three or four bottles of beer, but I didn't get out the grapefruit wine until it could be well iced.

Diz and two girls drove up to the cabin, and I proudly announced, "Tonight, you're not drinking home brew! I've got something really special!"

I brought in a bottle and opened it. It fizzed all over the place just like champagne. I hollered—no, ordered—"Catch it! Don't waste it!"

After everybody had a few sips, Diz squealed, "God

Aw'mighty, where'd you get this?"

"Made it."

"I didn't know you knew how to make champagne. How'd you know how to make it?"

"By the grace of God, I figured it out, and I just made it up!"

Diz had a pretty woman for me that night. I mean she was a pretty one, a woman from Okena who had recently been divorced. All three of them got about half soused and then went home.

About two o'clock the next afternoon, while I was on duty as a lifeguard, I was sitting in my tower. It was hot. I mean it was really HOT! Five women came up in a car and one of them who was with me at Camp Nasty the night before, called out, "Newt, come over here!"

When I got to their car, the women demanded, "Take us down to your cabin."

"What do you want to go to my cabin for? That place is nasty down there," I said.

They begged, "You've got something there we want! Every woman in Okena knows about that—that champagne you made!"

"God, Aw'mighty, you'll get me in trouble! No, I can't take you down there today. You'd better get with Diz and get him to take you out there, one at a time!"

Do you think I'd go down there with five women? Ain't no way in the world you'll get me down there!

From then on I was pestered about my brew and my frothy wine. I never made any more. The process was tedious and took too long, but I got to be well known around Okena for being able to turn grapefruit juice into sparkling wine.

I always remembered Mr. Ed Masters who brought

the grapefruit. He was a fine old gentleman!

21

"MR. NEWT! MR. NEWT!! Come over here! I wanna show you something," Willie Marsh said.

Willie Marsh was a pilot on the glass-bottom boat docked at the pier.

The whole sky was covered with clouds and it was drizzling rain, so there were not many paying customers at Sparklin' Waters, even though it was a Sunday afternoon. In fact, there was no one on Willie's boat except him. The weather was cool, but I was very comfortable in my monkey suit. It was late April, so school was still in session, and the crowds of tourists had not yet begun to come in. A single glass-bottom boat was downriver. There was one other tied to the loading pier, and three were lashed to the pier over by the maintenance shop, already tied down for the night. I was just roaming around the docks looking for customers. Willie got my attention. I climbed on board his launch at the far end of the passenger loading dock.

"What you got, Willie?" I asked after I untied the lines coupling the boat to the dock and pushed away the gangplank.

Willie pulled in the loose lines and fired up only one engine before he backed away from the dock. He steered the launch down the river only a few feet from the pier and pointed toward the floor of the river with his finger.

He yelled, "Look down there!"

When I first glanced toward the bottom of the river in the direction Willie was pointing, I didn't see anything unusual, but Willie insisted, "See that white thing down there. It looks like a giant tooth," he said.

I stared at the bottom of the river for almost a minute before I finally saw what Willie was pointing to. It did look like a tooth, a gigantic one. It was fifty feet below the surface. Now, I could see it, plain as day.

I pulled off my shoes and slipped out of my monkey suit. I still had on my bathing suit, which I rolled down so that it just covered my privates. Without thinking, I dived into the river from the bow of the glass-bottom boat. My dive was directly toward that thing that Willie was pointing to, but my angle of vision was skewed by the water, and that along with the flow of water in the river caused me to overshoot whatever it was by at least four feet. The current of the river was too fast for me to swim upstream to get to the object in the time I could stay down. I planted my feet against the river bottom and jumped back toward the surface. I climbed back onto Willie's launch and made another dive, a few feet upstream. This time, the current carried me directly to the thing we were looking at. I was able to grab onto the hard, white mass and hold on while I wiped away the sand. It was a tooth just like Willie thought, and there were lots of other skeletal bones buried in the sand around it. In the two-and-a-half minutes I was able to hold my breath, I pulled out the tooth and lugged it to the surface. Out of the water, it weighed nine pounds. I was able to pull that whole thing up to the surface without any diving gear or any other equipment. I made two more dives but was not able to pull out any of the other skeletal parts in the time I could stay down.

    The next morning, I ran over to the maintenance shop and found Mr. Ritchey. Together, we rounded up an old helmet which was used as a gas mask during World War I. We attached a rubber vest to it. One of my fellow workers was fitted with the diving equipment we concocted. He dived down toward the skeleton but was not able to get to the bottom of the river. When he failed to reach the target, I suited up, but I also fizzled. Whenever I leaned forward, the helmet filled with water and I was not able to see. We finally gave up trying to make dives using a homemade suit.

    Shorty Mayson heard about our struggles at pulling the skeleton from the bottom of the river and our efforts at putting together a homemade diving suit, and he came over to see what was going on. When he saw our flops, he called a deep-sea diving school in Tarpon Springs and made arrangements to send me over there to take a crash course to qualify as a professional diver. After two days of training, they put me on a launch and took me out into the gulf where a sponge diver helped me put on the diving gear. It was a big rubberized jacket with belts pulled tightly around my wrists. At the neck, the suit had rows of holes six inches apart. The suit was connected to two bronze plates by wing nuts to keep water from seeping in. A huge metal helmet with glass windows was fitted over the shoulders and attached. The bottom of the helmet rotated to seal it. Air was pumped into the diving suit and helmet, the amount of air being regulated by the exhaust valve in the back. If the valve opening pressure was too high, too much air collected in the suit, and the diver got a quick ride to the surface, feet first. The helmet filled with water, and deck hands had to stand the diver upright. That only happened once to any diver.

During the training, the beauty of the underwater plant and animal life in the Gulf of Mexico really excited me. There were big fish, little fish, porpoises, seaweed and lots, just lots of sponges. There were also some colorful jellyfish, but I tried to stay away from them.

After my course in deep-sea diving, I could hardly wait to get back to Sparklin' Waters to show off my expertise and the equipment. I was determined to pull up the pieces of skeleton I had seen in the bottom of the river. All the work crew who were not directly involved with the diving lined up along the riverbank to get a peek at what was going on. After digging away the sand, I was able to pull up a gigantic tusk. Then I kept going back until I got the whole skeleton. When we put it all together, we still didn't know what it was, but Shorty called the biology department at the University of Florida and was put in touch with a specialist who came over and inspected the bony parts. The shape of the teeth was like that of animals that fed on plants, the specialist said. The teeth looked like those of a cow or a horse, just a lot bigger. After putting all those things together, the biologist recognized that the skeleton was that of an *elepha columbi*, a prehistoric mammal, only a few of which were known to exist in Florida. The tusk was more than six feet long, but it crumbled when I was trying to get it up to the surface. We put all the skeletal parts together on the bank of the river and then collected all the parts of the tusk. We were finally able to glue it back together and put all of the parts of the skeleton together. We finally put it in a glass cage inside the ticket office at Sparklin' Waters. Unfortunately, the skeleton burned in a fire that wiped out the ticket office a few years later.

Details of the skeletal recovery were recorded in a note to Dr. William Scott of the faculty of the University of Florida on May 22, 1989.

While pulling up the skeletal parts from the floor of Sparklin' Waters, I had a chance to look inside the big cavern at the bottom of the river and found the pit just teeming with eels, crawfish and a lot of other marine animals. The eels were huge, three to four feet long. The crawfish had feelers which seemed to be as long as three feet, and their bodies were at least six inches in diameter.

Many years later, I was able to take home the diving helmet I used to get up the skeletal parts. I am keeping the helmet in a showcase in my home along with a lot of other reminders of my life at Sparklin' Waters and since then.

ROB ALLEN AND I ran across a newborn fawn out in the Okena National Forest. Its mother had been killed by a falling tree. We carried it back to the park and bottle fed it until it was able to eat on its own. The fawn hung around the lawn in front of the ticket office and restaurant at Sparklin' Waters, and it became a real pet. Visitors to the park liked to feed and pet it. He would walk up behind visitors sitting on the benches and nuzzle against them, especially if they had a lot of hair. Deer like to taste salt.

As the deer got older, he became a big tan buck with one antler that stood straight up. He became vicious and would rear up on his hind legs, often knocking down ladies or young children. He was getting to be a pest, and we were afraid that he might hurt somebody. We talked to a veterinarian who thought he had to be neutered if we were going to keep him running loose in the park. After the deer was fixed, he was once again gentle, a favorite of the kids.

22

"YOU'VE GOT TO be careful today. FBI agents are swarming all over the place," said Mr. Tracy.

The employees of Sparklin' Waters were startled by his announcement. They didn't know any reason why such high-ranking law enforcement officers should be snooping around their neighborhood. They didn't have any reason to think that what was going on around the park was out-of-the-ordinary. They kept working through the day just like they always had, but I could still hear some little rumblings. The employees just seemed to be uneasy. They kept whispering something back and forth to each other.

Throughout the town of Oklawaha, rumors spread that some of the people living there had connections with the underworld, but nobody was quite sure. Then word spread that a man named Dillinger had left the place where he was staying. Five carloads of folks thought to be thugs followed him toward the Northeast, I heard New York City. They got away before the FBI could raid the place.

The next morning about three o'clock, FBI agents stormed out of the Oklawaha Inn. They lined up along the shores of Lake Weir, near the home where Fred and Ma Barker rented a house. Fred Barker had a habit of sitting at a window on the top floor. Every so often, he took shots at ducks out on the lake. He used a fifty caliber machine gun—"Boom, boom, boom, boom!"

The sheriff watched to see where the shooting was coming from before the FBI joined the hunt.

About daylight the FBI agents surrounded the house and used a megaphone to order the Barkers to come out. Fred began firing again with both his machine gun and an automatic pistol. That started a real battle royal.

The FBI agents shot hundreds of rounds of ammunition into the house from every direction. Finally, after everything quieted down, they ordered John Willie Jones, an old black man who worked in the yard, to go into the house to see what was going on.

"Naw, sir, boss, I don't want to go in there. I can't go," he said.

The agent ordered, "You go up there now. I'm telling you to go."

John Willie finally sneaked up to the house and inside. He found Ma Barker lying on the floor with Fred lying beside her. Both were dead. Sam Pyles called a hearse to take their bodies to his funeral home in Okena.

That afternoon, Joey Leach, one of my buddies who worked at the funeral home, was swimming at Sparklin' Waters. He swam over to the lifeguard tower and in a loud whisper asked, "Newt, do you want to go see Ma and Fred Barker on the cooling board?"

I was always up for a challenge. I nodded. I was interested.

"Come up to the funeral parlor about eight thirty tonight. Wait until dark to come. I'll take you in, and we'll see them both."

Once I got inside the funeral home, Joey and I sneaked into the morgue and lifted the shroud covering

Ma Barker. There we saw a row of bullet holes, about four inches apart, forming a straight line from her left shoulder to her right hip. There were also holes in other parts of her body. When we raised the shroud covering Pa Baker, we could see holes all over his body too. That sight was awful. It caused me to feel sick to my stomach.

    This FBI raid broke up an underworld operation in what we thought was a quiet part of the world. It was a real surprise to everyone who worked at Sparklin' Waters.

23

"MR. PORTER?" THE deputy called out in a questioning manner to the short, lean, slightly graying man in a silk sport shirt and neatly pressed light blue trousers who came to the front door.

"Yeah."

"Mr. Rudolf Porter?" the deputy called out again, just to make sure.

"Yes, sir. That's me," Mr. Porter answered, assuring the deputy that he had the right name.

"I have a subpoena for you to appear in Judge Lefore's court at nine fifteen Friday morning."

"Really? What for?" It was now Mr. Porter's turn to be asking questions. But the deputy said only, "I don't know. I was just ordered to deliver this subpoena."

Several years after Sparklin' Waters became a real tourist attraction, a marsh along the riverfront not far from my cabin was filled in with sand. Nobody around the resort that I could find knew what was going on, but it was made into a nice sand beach with a new swimming area right out in front. The beach was about a hundred yards long. A spacious log cabin was built on the site. It looked like a dance hall. A sign above the log cabin bore the name "Paradise." After questioning a bunch of people around town, I found out that the whole project was being sponsored by Mr. Rudolf Porter, a wealthy lime-rock miner from Dunnellon, who was planning to compete with Sparklin' Waters.

Before Mr. Porter could start running glass-bottom boats from Paradise along Sparklin' Waters River, he had to build a road so travelers could get to his beach. To get there, he had to cut through a section of property owned by Sparklin' Waters. Signs were posted along the road, telling people how to get to Paradise.

Before he could open his new resort, Shorty charged him with trespassing. The case was heard in court in Okena, but no decision was reached. Shorty petitioned the Court of Appeals in New Orleans. That court came to the decision that, in effect, no company or business could cross another man's property to be in direct competition with the property owner. The order continued on, recommending that the part of the road passing through land owned by Sparklin' Waters be closed.

The court order was handed down about five o'clock one afternoon. As luck would have it, *Trouble in Paradise* was showing at the picture show in Okena. It's funny how the two events occurred at the same time! I'll never forget it.

The next morning, Shorty ordered every man who worked at Sparklin' Waters to come back to work at eight o'clock that night. He said to wear warm clothes and high-top shoes and be sure to bundle up because the temperature might be dropping.

When the employees got back to Sparklin' Waters, a big truck filled with rolls of barbed wire, staples, hammers, post hole diggers, shovels, axes, and saws was standing on the road to Paradise, just where it entered the land owned by Sparklin' Waters.

Workers were told to go over to the road, where each

man was assigned a job. They spent the night putting up fence posts and stringing barbed wire, crisscrossing the road time and time again. Even though it was the nighttime and dark, the work went on. Within hours, tractors and heavy equipment used to build roadways wrecked the road to Paradise.

Then a gate was put up blocking the entrance. A sign proclaimed, "Entrance into Paradise must be approved by an official at Sparklin' Waters."

Mr. Porter's daughter, who was an actress, walked up to the gate with a big sign still strapped to her back, which read, "Ride the glass-bottom boats from Paradise."

Shorty's night work made that impossible.

Paradise Club was later sold to Sparklin' Waters for chicken feed.

24

"I ALWAYS TOLD my people that whenever they are unhappy with the people they're working with, they should leave," said Mack Tracy.

"That's just what I plan to do. You owe me pay for a week's work and pay for a week's vacation," I replied.

Mr. Mack Tracy's recommendation to me and my response occurred when I complained that I was not able to work out a satisfactory arrangement with Diz Thomas for his taking some vacation time. Mr. Tracy looked me right in the eye when he gave his advice. His counsel was just what I expected. I don't know what I would have done if he had said something else. His recommendation came at a time when I had already decided to leave. I had just gotten word that my Daddy had died suddenly, and Mama needed me to help on the farms.

Before those things came about almost at the same time, I had never thought about leaving Sparklin' Waters. That was my home. I thought that I was a part of the company. But I still dreamed of going back to my roots. I just couldn't as long as Daddy was there.

It was a couple of months before that when Diz was invited to spend a week in Miami. He asked me to work a double shift to cover for him. I jumped at the chance, but I asked him to return the favor when it was time for me to take my vacation. I thought we had an agreement, but when the time came, he turned me down. That made me mad. He and I worked together for years, and I trusted him. I thought he was my friend, but he left me flat.

In the meantime. I had a call about the farm back in Ashburn, so I decided it was time for me to go back there. When I complained to Mr. Tracy and found no support, I was sittin' on "go."

Before that, I spent seven years at Sparklin' Waters. They were the best years of my life up to that time. When I was given my first assignment, the job of tearing down the dilapidated old shooting gallery, I knew I had to prove myself. When I was assigned to cut the grass along the riverbank, I still had to prove myself, but at least I had a way of making enough money to feed myself. After that, I had a home. One after another, my bosses took an interest in me. They looked after me. They taught me. Through the years, I felt more and more a part of the team. Sparklin' Waters was my home. My life there was exciting. I had new adventures every day. Stupendous experiences just seemed to pop up. Every day was different.

As my life there came to an end, I was thankful for my bosses, each of whom I held in high regard. I admired them. I was thankful for their friendship, and I felt I owed them everything. Having only five-and-a-half years of formal education, I felt that I was taught a lot of practical things, things that would make up for my lack of formal schooling. Those teachings would provide me with a way to make a living, no matter what my future held.

Besides my close ties with my bosses, my aunt, my cousins, and my fellow workers at Sparklin' Waters, my job led me to meet a lot of interesting people, some of whom were really prominent. My associations with the guests and the local people that I met could not have been more profitable. Each person had his or her own influence. The effects of some of them have stayed with me for all my life. They might have changed my personality. Even as the years go by, I think back and realize that my experiences at Sparklin' Waters were really valuable.

When I told Mr. Tracy that I was ready to leave, we climbed the stairs to his office where he wrote a check for two weeks' wages. He handed it to me, and I cashed it in the ticket office. I walked out to the sand beach and pushed my canoe onto the water and climbed in. I poled down to Camp Nasty and tied the canoe to the jetty. I walked up to the shack and collected everything I could carry. Then I walked to the Okena bus station where I bought a one-way bus ticket to Ashburn, Ga. I was going home. It almost seemed like a dream, a long dream but a good one.

About the Author

Marshall Allen, Jr., M.D., is a retired neurosurgeon and the former Chief of Neurosurgery for the Medical College of Georgia. He and his wife reside in Augusta, Ga. *The Naked Bus Driver* is his first novel about Newt Sexton, and he is now working on a third one.

To order copies of *Newt in the World of Tarzan, The Naked Bus Driver* or other titles from Harbor House, visit our Web site:

www.harborhousebooks.com